GUIDE

NSPECT
nd CERT

cluding periodic reporting - 8th Edition

updated to

18TH EDITION

BS 7671:2018

RTSURE

rtsure LLP is a limited liability partnership between the Electrical Safety First and the Electrical Contractors' sociation (ECA) registered in England and Wales. Its brands include NICEIC the UK's leading voluntary gulatory body for the electrical contracting industry, which has been assessing the electrical competence of ectricians for over fifty years and currently maintains a roll of over 36,000 registered contractors. It also includes ECSA, who provide inspection, assessment and certification services to contractors working with electrical and newable technology installations in private dwellings.

ertsure products are delivered through the NICEIC and ELECSA brands. The products include, amongst others:

The NICEIC Approved Contractor Scheme, a scheme for electrical contractors undertaking design, installation, commissioning and maintenance of electrical installations to *BS 7671*.

The ELECSA and NICEIC Part P Domestic Installer Scheme, which registers installers to Full Scope or Defined Competence for domestic electrical installation work in accordance with Part P of the Building Regulations in England and Wales.

The Microgeneration Certification Scheme (MCS), developed to assess businesses installing microgeneration technology and is designed to protect the consumer.

The Green Deal Scheme, for energy saving improvements to dwellings and other buildings.

LECTRICAL SAFETY FIRST

lectrical Safety First (formerly the Electrical Safety Council) is an independent charity committed to reducing eaths and injuries through electrical accidents at home and at work.

lectrical Safety First is supported by all sectors of the electrical industry, approvals and research bodies, onsumer interest organisations, the electricity distribution industry, professional institutes and institutions, egulatory bodies, trade and industry associations and federations, trade unions and local and central government.

HE ELECTRICAL CONTRACTORS' ASSOCIATION

he Electrical Contractors' Association (ECA) is the leading trade association for contractors involved in electrical engineering projects in England, Northern Ireland and Wales.

ublished by:

ertsure LLP, Warwick House, Houghton Hall Park, Houghton Regis, Dunstable, Bedfordshire LU5 5ZX

el: 0333 015 6626
Vebsites: www.certsure.com www.niceic.com www.elecsa.co.uk www.shop.niceic.com

SBN: 978-1-906091-90-3 EAN: 9781906091903 Sap code: PNICITC18

ıblished July 2018

CONTENTS

CHAPTER 1
INTRODUCTION

1.1 Aims and objectives

This Guide is intended to complement Part 6 of *BS 7671: 2018* and covers the general requirements relating to the inspection and testing of electrical installations forming part of TN-C-S, TN-S and TT systems. In particular, the Guide provides practical advice and guidance on completion of the following:

- NICEIC and ELECSA electrical installation certificates, including the minor works certificate,
- NICEIC electrical installation condition reports,
- schedules of inspections and test results for certificates and reports, and
- performing the range of test methods required and verifying the results.

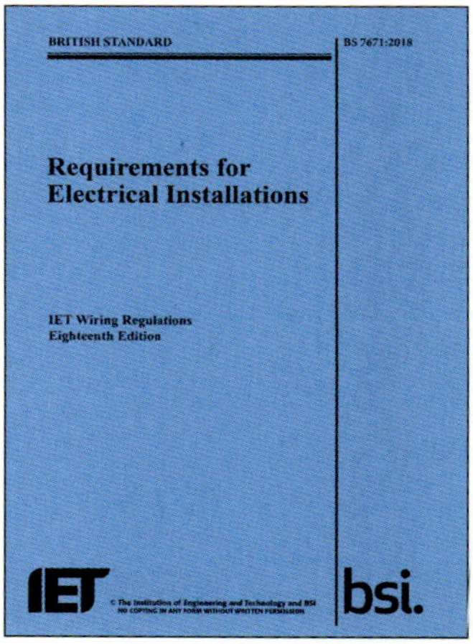

Fig 1.1 *BS 7671: 2018* - Requirements for Electrical Installations

Note: This Guide does not cover particular requirements relating to TN-C and IT systems, which are not common in the UK.

Even the most experienced electrical engineers and electricians can be expected to make the occasional mistake when carrying out electrical installation work. This is one of the main reasons why every new electrical installation, and every addition or alteration to an existing installation, must be thoroughly inspected and tested, and any defects or omissions found made good, before the installation, or the addition or alteration, is put into service. The other key reason for inspection and testing is to provide a record of the condition of a new installation when it is first put into service, or when an altered or extended installation is restored to service.

All electrical equipment deteriorates with age, as well as with wear and tear from use.Every electrical installation therefore needs to be inspected and tested at appropriate intervals during its lifetime to establish that its condition is such that, subject to the completion of any necessary remedial work, the installation is safe to remain in service at least until the next inspection is due. Correct compilation of certificates and reports provide the persons responsible for the safety of electrical installations (including contractors, owners and users) with an important record of the condition of electrical installations at the time they were inspected and tested.

Additionally, certificates and reports also provide an essential basis for subsequent inspection and testing, without which a degree of costly exploratory work might be necessary on each occasion. In the event of injury or fire alleged to have been caused by an electrical installation, certificates and reports will provide documentary evidence to help demonstrate that, in the opinion of skilled persons, competent in such work, the installation had been installed and subsequently maintained to a satisfactory standard of safety.

1.2 Scope

This Guide is not intended to instruct untrained and inexperienced persons to undertake the inspection and testing of electrical installations. It assumes that all persons proposing to undertake such work have already acquired the necessary knowledge, understanding and skill, and are properly equipped to undertake such work without putting themselves and others at risk.

Persons who consider that they are not fully skilled (electrically) to undertake the inspection and testing of electrical installations without direct supervision should, as a first step, complete an appropriate training course. For details on training courses at NICEIC and ELECSA, refer to: **www.shop.niceic.com/training**

1.3 Arrangement of the Guide

The main theme of this Guide is the preparation of NICEIC and ELECSA and the related uncontrolled (green) versions of the principal forms of certification and reporting used in the electrical industry. All the relevant details and declarations relating to the safety of an electrical installation are intended to be recorded on these forms, which consist of:

- The Electrical Installation Certificate (EIC) - intended for new installation work, including additions and / or alterations to existing installations.
- The Domestic Electrical Installation Certificate (DEIC) - intended for new installation work, including additions and / or alterations to existing installations not exceeding 100 A single-phase, (a house, flat or similar).
- The Minor Electrical Installation Works Certificate (MEIWC) - intended for minor installation work that does not include the provision of a new circuit.
- The Electrical Installation Condition Report (EICR) - intended for reporting on the condition of an existing installation.

For guidance on completion of the 'Domestic Electrical Installation Condition Report' - intended for reporting on the condition of an existing installation in a single dwelling (a house, flat or similar), refer to the NICEIC and ELECSA publication; *Domestic Periodic Inspection, Testing and Reporting.*

The particular versions of the forms on which this book is based are those published by NICEIC for use by Approved Contractors and Domestic Installers and those published for use by ELECSA registered contractors. However, the guidance is equally applicable to other forms of certification and reporting based on the model forms given in Appendix 6 of *BS 7671: 2018,* especially the green versions which are for use where contractors registered with NICEIC and ELECSA undertake work outside the scope of their registration.

Some of the inspection and testing procedures required for periodic inspection reporting are common to those required to verify new installation work. To avoid duplication, such procedures are covered in the book only once, the reader being referred to another chapter where appropriate.

NICEIC APPROVED CONTRACTOR — RED (badged) certificates and reports available for use by NICEIC Approved Contractors

NICEIC DOMESTIC INSTALLER — RED (badged) certificates available for use by NICEIC Domestic Installers

ELECSA — BLUE (badged) certificates available for use by ELECSA registered contractors

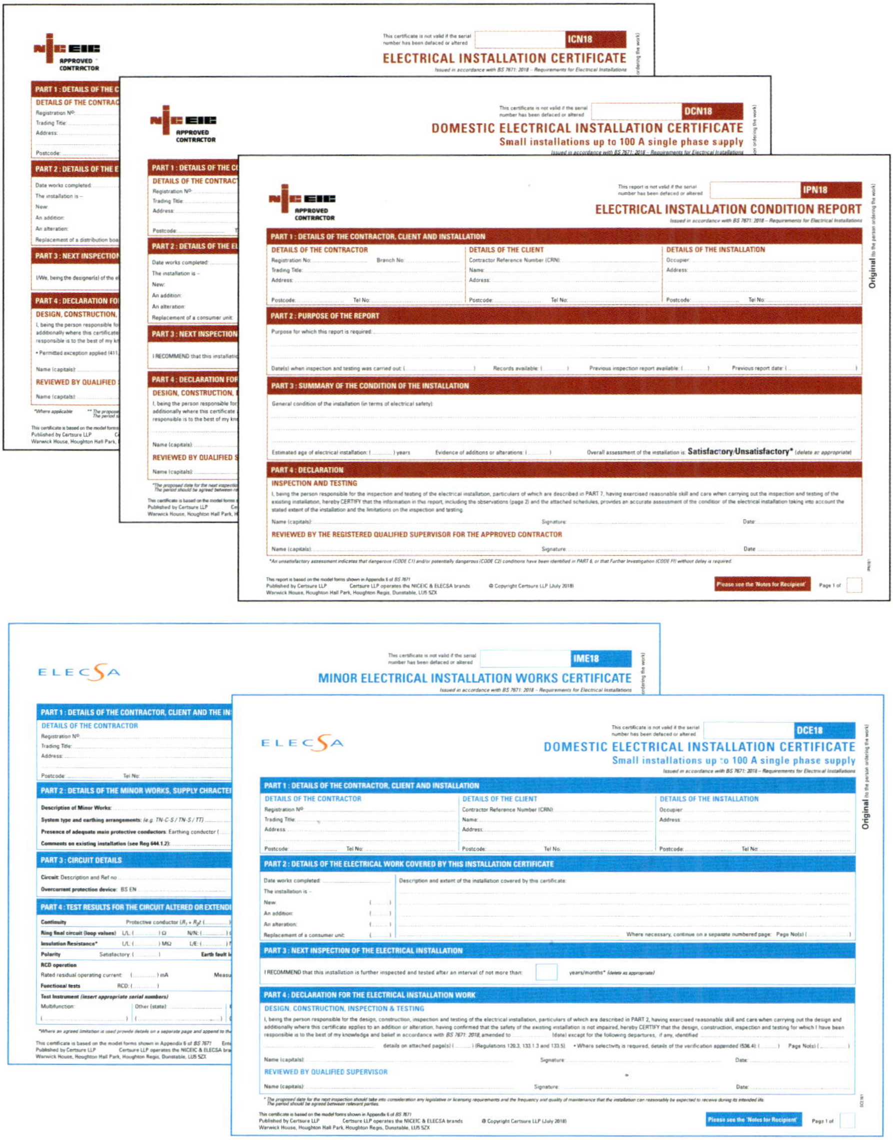

Fig 1.3 NICEIC and ELECSA certificates and reports

1.4 The purpose of inspection, testing and certification or reporting

The fundamental reason for inspecting and testing an electrical installation is to determine whether the installation work is safe to be put into service, or whether an existing installation is safe to remain in service until the next inspection is due.

Initial verification is required to confirm that a new installation or an addition or alteration to an existing installation including the replacement of a distribution board or consumer unit, is safe to be put into service. Verification includes inspection and testing during construction and on completion, and comparison of the results with relevant criteria, and with any particular electrical safety requirements detailed in the design information, to confirm that the requirements of *BS 7671* have been met (Regulation 641.3 refers).

Certification provides an assurance to the person ordering the work that the electrical installation has been designed and installed in accordance with *BS 7671* (Regulation 641.1 refers). Where the work is subject to a departure from *BS 7671*, the certificate requires the designer to provide confirmation that the departure will not result in a lesser degree of safety than would have been achieved by full compliance with the Standard (Regulation 120.3 refers). Any defects or omissions found during the process of initial verification must be corrected (and inspection and tests repeated as necessary) before the certificate is issued.

For an addition or alteration to a circuit and not extending to the provision of a new circuit, a Minor Electrical Installation Works Certificate may be used instead of an Electrical Installation Certificate or a Domestic Electrical Installation Certificate.

Having been certificated and put into service, generally electrical installations need to be inspected and tested at appropriate intervals throughout their lifetime. On each occasion, details of the condition of the installation should be recorded in an Electrical Installation Condition Report for the benefit of the person ordering the inspection, and of the persons subsequently involved in remedial work, additions or alterations, or further inspections.

1.5 NICEIC and ELECSA certificates and reports

NICEIC and ELECSA certificates and report forms are accountable documents and as such, unused copies should be kept secure by the Qualified Supervisor and records of each certificate or report issued should be made on the record sheet provided.

A unique serial number (including prefix) is printed on the original and duplicate of each certificate and report form published for use by contractors registered with NICEIC or ELECSA. The number allows the origin of the certificate or report to be traced by the respective registration body in the event of a query as to its authenticity. For this reason, only the organisation supplied with a certificate or report from NICEIC or ELECSA is authorised to issue the form. If the serial number has been erased or altered in any way, the certificate or report is deemed invalid.

NICEIC and ELECSA strongly prefer their registered contractors to issue certificates and report forms supplied by their registration body as this provides a measure of safety assurance and confidence to the recipients.

Uncontrolled versions of NICEIC and ELECSA certificates and reports, coloured green and not serially-numbered, are available for use by NICEIC or ELECSA registered contractors undertaking work outside the scope of their registration or those preparing for registration.

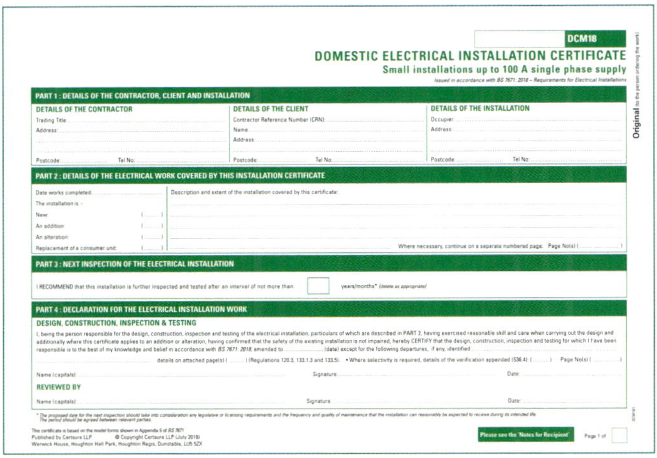

Fig 1.5. Uncontrolled green version of the Domestic Electrical Installation Certificate

NICEIC and ELECSA online certification service

In addition to printed pads of NICEIC and ELECSA certificates and forms, and those made available for use with computer software packages, all principal certificates and report forms published by NICEIC and ELECSA may be completed using our online certification service.

The NICEIC and ELECSA online service allows contractors to certify and notify work using the same system. Certificates and forms may be issued and notified on-site, in the office or at home using a PC, tablet or smartphone and the records, including any digital images or additional documents relating to the work, securely stored in the 'Cloud'. For information about online certification refer to the appropriate website address.

www.niceiconline.com or **www.elecsaonline.com**.

It should be noted that irrespective of the method used to compile certificates and reports, it remains the responsibility of the person issuing the document to ensure that it has been compiled correctly before being issued. NICEIC and ELECSA certificates and reports must be reviewed by a Qualified Supervisor for the particular contractor, who in signing the document is confirming that it has been correctly compiled.

Regardless of the type of certificate issued or the media used, written or electronic, the authenticity and integrity of certificates or their copies, shall be verified by a reliable process or method (Regulation 644.4.202 refers).

1.6 The inspector

Throughout this book, the term 'inspector' is used to describe a person responsible for inspecting and testing an electrical installation. Inspection and testing of new or existing electrical installations must be undertaken by skilled persons competent in such work (Regulations 641.6 and 651.5 refer).

If the inspector is a skilled person competent in such work and takes all the necessary safety precautions, the work of inspection and testing should not create any danger to persons or livestock, or cause damage to property or equipment (Regulations 641.4 and 651.3 refer).

NICEIC and ELECSA consider that, to undertake the inspection and testing of an electrical installation, persons should as a minimum:

- Have sufficient knowledge and experience of electrical installation matters to avoid danger to themselves and to others.
- Be familiar with, and understand, the requirements of *BS 7671*, including those relating to inspection, testing, certification and reporting.
- Have a sound knowledge of the particular type of installation to be inspected and tested.
- Have sufficient information about the function and construction of the installation to allow them to proceed in safety.

It is not sufficient for the inspector simply to record the results of inspection and testing on a certificate or report. The inspector should consider and assess the results and compare them with the relevant criteria to determine whether they are as expected for the particular type of installation, and that they are consistent with installation work complying with all the relevant requirements of *BS 7671* (Regulation 641.3 refers).

Generally persons undertaking periodic inspection, testing and reporting need to have above-average knowledge and experience of electrical installation matters to enable them to safely and accurately assess the condition of an existing electrical installation, especially when they do not have access to the design information relating to that installation.

1.7 Safe working procedure

All persons carrying out inspection and testing should employ safe working procedures and be familiar with the relevant statutory provisions applicable to the work. Statutory requirements are listed in Appendix 2 of *BS 7671*. For example, the inspector should be familiar with the requirements of building regulations and where appropriate, such as for places of entertainment, the inspector should take account of the requirements of the licensing authority, as required by Regulation 115.1.

All persons carrying out inspection and testing must comply with the statutory requirements of 'The Electricity at Work Regulations 1989 (EWR)'. It should be noted that whilst *BS 7671* is a non-statutory document, it may be used in a court of law as evidence of compliance with a statutory requirement (Regulation 114.1 refers). For example, compliance with *BS 7671* it is likely to achieve compliance with relevant aspects of the statutory requirements of 'The Electricity at Work Regulations'.

The Health and Safety Executive (HSE) guidance document HSR25, 'Electricity at Work Regulations 1989 - Guidance on Regulations' provides technical and legal guidance on the EWR which is intended to support dutyholders to meet the requirements of the EWR (including EWR 1991, Northern Ireland).

Fig 1.7.1 Electricity at Work Regulations 1989 - Guidance on Regulations HSR25 (Third edition)

Note: HSR25 is available to download from: **www.hse.gov.uk/pUbns/priced/hsr25**

Persons undertaking inspection and testing should pay particular attention to Regulation 14 of the EWR; 'working on or near live conductors':

> 'No person shall be engaged in any work activity on or so near any live conductor (other than one suitably covered with insulating material so as to prevent danger) that danger may arise unless:
>
> - it is unreasonable in all the circumstances for it to be dead; and
> - it is reasonable in all the circumstances for him/her to be at work on or near it while it is live; and
> - suitable precautions (including where necessary the provision of suitable protective equipment) are taken to prevent injury.'

In view of the risks associated with 'live' testing, wherever practicable the information required should be determined by an alternative method, such as by enquiry or calculation. Where live testing is performed it should only be undertaken by persons competent in such work, employing suitable tools, including insulated tools, instruments and test probes.

Regulation 12: of the EWR requires that where necessary to prevent danger, suitable means (including, where appropriate, methods of identifying circuits) are made available:

- for cutting off the supply of electrical energy to any electrical equipment, and
- for the isolation of electrical equipment.

Isolation is not simply the operation of an isolating device. It is a procedure, the objective of which is to ensure that the supply is secured for reasons of safety. Details of the isolation procedure are provided in Appendix A of this Guide:

Note: In some circumstances switching off or isolating equipment may not be practicable because the source of electrical energy is an integral part of the equipment, such as the solar panels of a photovoltaic system. Where this is the case all reasonable safety precautions should be observed.

HSG85 'Electricity at Work safe working practices' gives detailed guidance on devising safe working practices for people who carry out work on or near electrical equipment the guidance is available at: **www.hse.gov.uk/pUbns/priced/hsg85.pdf**

Fig 1.7.2 HSG85 Electricity at Work, Safe working practices

1.8 Test equipment

Test instruments and their associated test leads must be suitably constructed and fit for the purpose for which they are to be used. Such equipment should preferably meet the safety requirements set out in *BS EN 61010* and the performance standards detailed in the relevant Part(s) of *BS EN 61557* (Regulation 643.1 refers). HSE Guidance Note GS38: 'Electrical test equipment for use on low voltage electrical systems', provides guidance for those who use electrical test equipment on low voltage electrical systems and equipment (see Appendix B of this Guide).

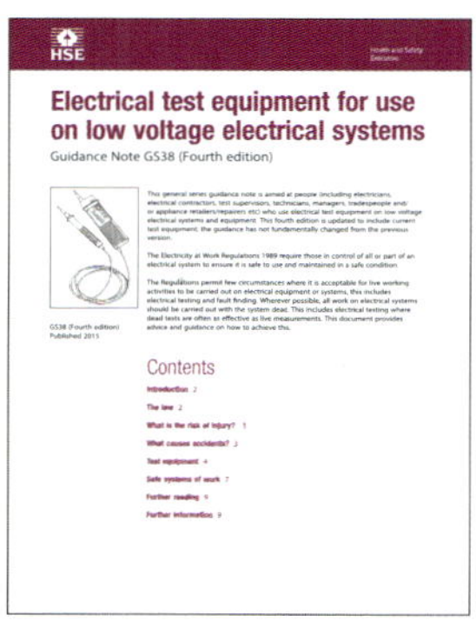

Fig 1.8.1 HSE Guidance Note GS38 - Electrical test equipment for use by electricians

Note: GS38 is available for download at **www.hse.gov.uk/pUbns/priced/gs38.pdf**

CHAPTER 2

THE ELECTRICAL INSTALLATION CERTIFICATE (AND THE DOMESTIC ELECTRICAL INSTALLATION CERTIFICATE)

(i) General

This Chapter looks at the completion of NICEIC and ELECSA Electrical Installation Certificates and provides guidance that is applicable to:

- The red NICEIC Electrical Installation Certificate (EIC); only an NICEIC Approved Contractor or Conforming Body responsible for the construction of the electrical installation is authorised to issue this certificate.
- The red NICEIC (Approved Contractor) Domestic Installation Certificate (DEIC) - Small installations up to 100 A single-phase supply; only an NICEIC Approved Contractor is authorised to issue this certificate.
- The red NICEIC (Domestic Installer) Domestic Installation Certificate (DEIC) - Small installations up to 100 A single-phase supply; only an NICEIC registered Domestic Installer is authorised to issue this certificate.
- The blue ELECSA Domestic Installation Certificate (DEIC) - Small installations up to 100 A single-phase supply; only an ELECSA registered contractor is authorised to issue this certificate.

For those not authorised to issue the certificates listed above, similar uncontrolled (green) certificates are made available from NICEIC and ELECSA and are intended for use by electrical contractors not registered with NICEIC or ELECSA, such as those preparing for registration, or those contractors registered with NICEIC or ELECSA working outside the scope of their registration.

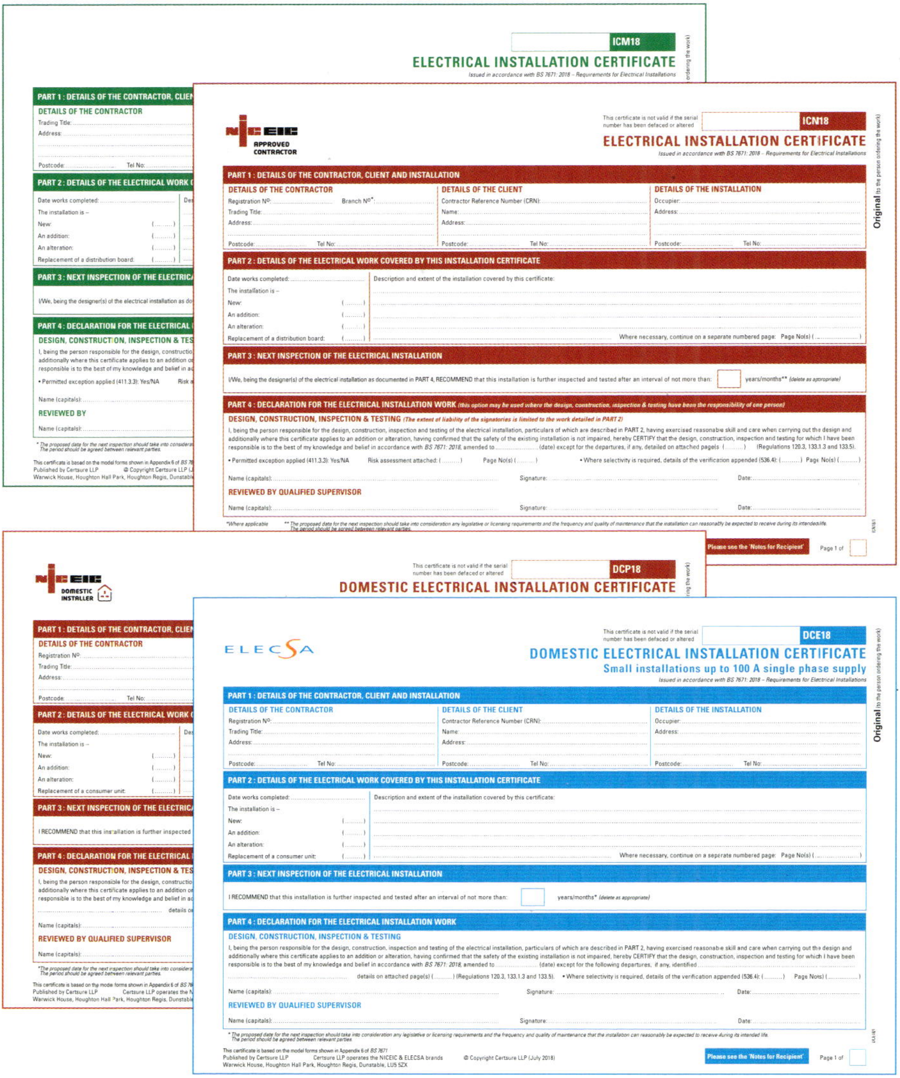

Fig (i) NICEIC and ELECSA Electrical Installation Certificates

(ii) Requirements

As required by Regulation 644.1 an Electrical Installation Certificate (EIC) must be issued for all new installation work and for all additions and alterations to existing installations, including the replacement of a distribution board or consumer unit. The certificate should be issued to the person ordering the work as a formal assurance from those responsible for the design, construction, and inspection and testing of the installation that the work complies with the requirements of *BS 7671*, as amended (Regulation 644.4 refers). A separate certificate must be issued for each distinct electrical installation, and should include the details of the extent of the installation covered and a record of the inspections and the results of the testing (Regulation 644.3 refers). Any defects or omissions identified during the initial verification of the electrical work should be corrected before the certificate is issued.

Subject to certain limitations, an NICEIC or ELECSA Domestic Electrical Installation Certificate (DEIC) may be used for the certification of electrical installation work typically relating to a single dwelling (house or individual flat), as long as the installation work falls within the scope of that certificate. For work outside the scope of the DEIC an EIC must be issued. For those not authorised to issue the NICEIC, Electrical Installation Certificate, a green uncontrolled version is available. However, irrespective of the type of certificate used, it should be noted that the issue of an electrical certificate is a declaration of electrical safety and as such it should not be withheld for contractual reasons, especially if the electrical installation is available for use.

NICEIC and ELECSA certificates must only be issued by those authorised to do so, and should never be issued to cover another contractor's installation work.

Where the electrical work does not extend to the provision of a new circuit, a Minor Electrical Installation Works Certificate (MEIWC) may be used. Guidance on the completion of the MEIWC is given in Chapter 3.

(iii) Additions or alterations

For an addition or an alteration it should be verified that the work complies with *BS 7671*, as amended, and does not impair the safety of the existing installation (Regulation 641.5 refers). Any defects observed on the existing installation when undertaking the addition or alteration that may give rise to danger should be recorded on the certificate, under 'Comments on the existing installation' whereas any defect that affects the safety of the work carried out must be corrected before the certificate is issued (Regulation 644.1.2 refers).

Before an addition or alteration is made to an existing installation, the designer is required to ascertain that the rating and condition of any existing equipment, including that of the distributor will be adequate for the altered circumstances, and that the earthing and protective bonding arrangements, if necessary for the protective measure applied for the safety of the addition or alteration are also adequate (Regulation 132.16 refers).

Where a contractor discovers the existence of a dangerous or potentially dangerous situation in the existing installation (such as the absence of earthing or protective bonding where the protective measure for fault protection is Automatic Disconnection of Supply (ADS), the new work should not proceed and the client should be advised immediately, preferably in writing, to satisfy the duties imposed by the *Electricity at Work Regulations 1989* (or 1991 for Northern Ireland). This may be achieved by the issue of an Electrical Danger Notification form to the client.

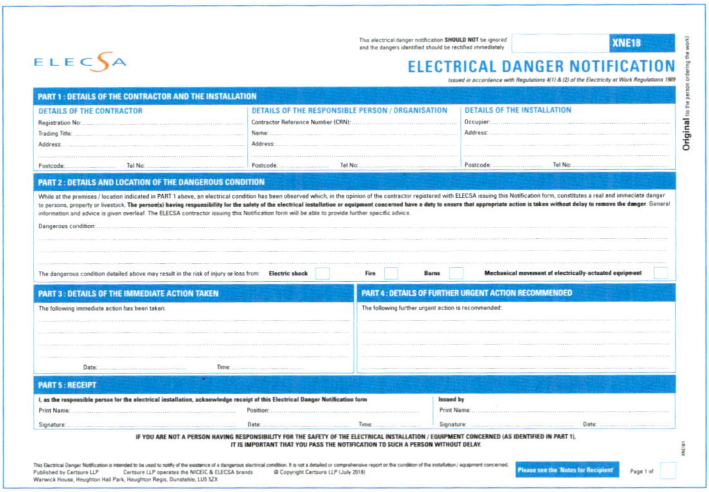

ELECSA

This electrical danger notification **SHOULD NOT** be ignored and the dangers identified should be rectified immediately

XNE18

ELECTRICAL DANGER NOTIFICATION

Issued in accordance with Regulations 4(1) & (2) of the Electricity at Work Regulations 1989

Original (to the person ordering the work)

PART 1 : DETAILS OF THE CONTRACTOR AND THE INSTALLATION

DETAILS OF THE CONTRACTOR	DETAILS OF THE RESPONSIBLE PERSON / ORGANISATION	DETAILS OF THE INSTALLATION
Registration No:	Contractor Reference Number (CRN):	Occupier:
Trading Title:	Name:	Address:
Address:	Address:	
Postcode: Tel No:	Postcode: Tel No:	Postcode: Tel No:

PART 2 : DETAILS AND LOCATION OF THE DANGEROUS CONDITION

While at the premises / location indicated in PART 1 above, an electrical condition has been observed which, in the opinion of the contractor registered with ELECSA issuing this Notification form, constitutes a real and immediate danger to persons, property or livestock. **The person(s) having responsibility for the safety of the electrical installation or equipment concerned have a duty to ensure that appropriate action is taken without delay to remove the danger.** General information and advice is given overleaf. The ELECSA contractor issuing this Notification form will be able to provide further specific advice.

Dangerous condition:

The dangerous condition detailed above may result in the risk of injury or loss from: **Electric shock** ☐ **Fire** ☐ **Burns** ☐ **Mechanical movement of electrically-actuated equipment** ☐

PART 3 : DETAILS OF THE IMMEDIATE ACTION TAKEN	PART 4 : DETAILS OF FURTHER URGENT ACTION RECOMMENDED
The following immediate action has been taken:	The following further urgent action is recommended:
Date: Time:	

PART 5 : RECEIPT

I, as the responsible person for the electrical installation, acknowledge receipt of this Electrical Danger Notification form	**Issued by**
Print Name: Position:	Print Name:
Signature: Date: Time:	Signature: Date:

IF YOU ARE NOT A PERSON HAVING RESPONSIBILITY FOR THE SAFETY OF THE ELECTRICAL INSTALLATION / EQUIPMENT CONCERNED (AS IDENTIFIED IN PART 1), IT IS IMPORTANT THAT YOU PASS THE NOTIFICATION TO SUCH A PERSON WITHOUT DELAY.

This Electrical Danger Notification is intended to be used to notify of the existence of a dangerous electrical condition. It is not a detailed or comprehensive report on the condition of the installation / equipment concerned.
Published by Certsure LLP Certsure LLP operates the NICEIC & ELECSA brands © Copyright Certsure LLP (July 2018)
Warwick House, Houghton Hall Park, Houghton Regis, Dunstable, LU5 5ZX

Please see the 'Notes for Recipient' Page 1 of

XNE18/1

Fig (ii) Example of an Electrical Danger Notification form

Where the electrical work to which the certificate relates includes the installation of a fire detection/alarm system and/or an emergency lighting system (or a part of such systems) in accordance with British Standards *BS 5839* and *BS 5266* respectively, the appropriate certificate or certificates as prescribed by those standards should accompany the electrical certificate.

(iv) Continuation sheets

Depending on the number and type of circuits in the installation additional numbered pages may need to be appended to the electrical certificate. For installations having more than one consumer unit or distribution board or more circuits than can be recorded on the combined 'Schedule of Circuit Details and Test Results', one or more additional schedules will be required.

Continuation sheets designed for such purposes are available separately from NICEIC and ELECSA (see Chapter 5).

2.0 Compilation of the certificate

Irrespective of the method used to compile NICEIC and ELECSA certificates, it remains the responsibility of the compiler to ensure that the information provided on the certificate is factual, and that the electrical installation work to which the certificate relates is safe to be put into service.

The certificates should be completed by inserting the necessary text or value, or a 'tick' should be inserted to indicate that the task has been completed and the result was satisfactory. Where an item is not applicable to the work undertaken N/A' (Not Applicable) should be entered. Where additional information is required, the details should be provided on a separate numbered page(s) and attached to the certificate.

The remainder of this Chapter provides guidance on completing the NICEIC Electrical Installation Certificate (EIC) and the NICEIC and the ELECSA Domestic Electrical Installation Certificate (DEIC). Whilst the layout and most aspects of the EIC are common to the NICEIC or ELECSA DEIC, some of the parts are numbered differently. Therefore, where a particular part in the EIC is numbered differently to the DEIC it is identified in the text, and where a particular aspect or part being discussed is applicable to the EIC only, it is identified by the bracket '(EIC only)'.

DETAILS OF THE CONTRACTOR
Registration N^o^:.............................. Branch N^o^*:..............................
Trading Title:..
Address:..
....
Postcode:......................... Tel No:..

Fig 2.1.1 Details of the contractor

2.1 Part 1: Details of the contractor, client and installation

The details required here are straightforward and generic to all forms of NICEIC and ELECSA certificates. The contractor's registration number must be inserted and where applicable the branch number. As the registration and the branch number are not applicable to the green version of these certificates (or report forms) these items are not included on the green versions.

DETAILS OF THE CLIENT
Contractor Reference Number (CRN):...
Name:..
Address:...
....
Postcode:......................... Tel No:...

Fig 2.1.2 Details of the client

The details that should be recorded are the name and address of the person or organisation that engaged the contractor to carry out the installation work. Where, for example, the electrical contractor is engaged by a builder, the details entered should be those of the builder. In the case of lengthy name and address details, suitable abbreviations may be used if acceptable to the client.

The item entitled 'Contractor Reference number (CRN)' is provided so that the contractor can assign a particular reference to the job or client, as appropriate. Otherwise 'N/A' should be inserted in the space. Where a green version of a certificate is used, the CRN may be used as the reference number for any continuation sheets or additional pages that have to be attached to the certificate.

DETAILS OF THE INSTALLATION

Occupier: ....

Address: ....

....

....

Postcode: .... Tel No: ....

Fig 2.1.3 Details of the installation

It should be appreciated that the address recorded here may be different from the address of the person ordering the work (client). For example, the client may have several premises. Likewise, for a domestic installation, it may be a landlord who orders the work. Therefore, the full address of the premises where the installation work has been carried out should be recorded along with the name of the person or company that is resident at the address.

For work carried out at owner-occupied premises the same name and address should be entered in both sections.

2.2 Part 2: Details of the electrical work covered by this installation certificate

Along with the date the works were completed, the nature of the installation work undertaken should be indicated using the fields provided:

- **'New'** - This field should be ticked only if the whole installation has been installed as new, or if a complete rewire has been carried out.
- **'An addition'** - This field should be ticked if an existing installation has been modified by the addition of one or more new circuits.
- **'An alteration'** - This field should be ticked where one or more existing circuits have been modified or extended, or items of switchgear have been replaced.

- **'Replacement of a distribution board'** - This box should be used to indicate that the work has involved the replacement of a distribution board (or consumer unit).

Where appropriate, more than one category should be ticked. Where an item is not applicable, insert 'N/A'

Date works completed:..		Description and extent of the installation covered by this certificate:
The installation is –		....
New:	(.........)	....
An addition:	(.........)	....
An alteration:	(.........)	....
Replacement of a distribution board:	(.........)	....

Fig 2.2 Details of the electrical work covered by this installation certificate

Description and extent of the installation covered by this certificate

The extent of the installation covered by the certificate must be clearly defined. The certificate may apply to the whole of the electrical installation at the address given or a particular floor or area of the premises. In addition to defining the work undertaken, it may also be beneficial, especially in the case of additions or alterations, to identify work that is not covered by the certificate. For example, where particular circuits or systems are excluded from the agreed work, such as heating and control circuits or fire detection and alarm systems, it is advisable to record such exclusions on the certificate.

Where the EIC relates to a mobile or transportable unit or a leisure accommodation vehicle, details of the make and model of the vehicle/unit together with a unique means of identification, such as the vehicle identification number (VIN), should be recorded as part of the description of the installation.

If there is insufficient space in the box to give a clear description of the work and/or any exclusions to the work, the additional details should be provided on a separate numbered page and attached to the certificate. The additional page number should be recorded in the space provided and also in the section entitled, 'Schedules and Additional Pages'.

It should be appreciated that failure to clearly describe the extent of the work covered by the certificate could involve the contractor in unforeseen liabilities at a later date.

2.3 Part 3: Next inspection of the electrical installation

As required by Regulations 644.4 and 134.2.2, the certificate should contain a designer's recommendation for the period between initial verification and the first periodic inspection. The time interval recorded, in terms of years or months should take into consideration any legislative or licensing requirements and the frequency and quality of maintenance that the installation can reasonably be expected to receive during its intended life. In no circumstances should this section be left blank or 'N/A' (Not Applicable) be inserted. Guidance on periodic inspection is given in Chapter 4.

PART 3 : NEXT INSPECTION OF THE ELECTRICAL INSTALLATION

I/We, being the designer(s) of the electrical installation as documented in PART 4, RECOMMEND that t

Fig 2.3 Next inspection of the electrical installation

Note: In some premises periodic inspection and testing may be replaced by an adequate regime of continuous monitoring and maintenance (Regulation 652.2 refers).

2.4 Part 4: Declaration for the electrical installation work

This part of the certificate must be completed to confirm that the three elements of the installation work; the design, the construction and the inspection and testing have been carried out in accordance with *BS 7671: 2018*.

Where all three elements are the sole responsibility of the Approved Contractor issuing the EIC, the elements may be certified collectively by completion of the section headed: 'Design, Construction, Inspection and Testing'. This single declaration option is similar to the declaration used on the NICEIC or the ELECSA Domestic Electrical Installation Certificate.

Where the NICEIC Approved Contractor issuing the EIC has not been responsible for all aspects of the work, separate declarations must be provided in accordance with Regulation 644.4 by those responsible for:

- Design
- Construction
- Inspection and testing

Whichever declaration option is used, the date of the amendment to *BS 7671: 2018* current at the time the work was carried out should be stated in the space allocated in the declaration. The following information relates to the single declaration (refer to Fig 2.4.1).

2.4.1 Single declaration for design, construction, inspection and testing

Where the single declaration is used, the person signing the declaration is accepting responsibility for all three elements of the electrical installation work: design, construction and inspection and testing. Where this is the case, 'N/A' should be recorded in the boxes provided for the separate declarations.

It should be noted that where no other person is specifically responsible for the installation design, then the contractor will inevitably assume the role of the designer. The designer is responsible for all aspects of the installation design, such as the selection of protective devices, cables and equipment. This is the case even if the design work only involves the application of industry standards, such as for ring final circuits. In such cases the contractor will be responsible for the correct application of the industry standards and, consequently, for the installation design.

Departures

BS 7671 recognises that designers should not be restricted from specifying new technologies, equipment or systems simply because they are not covered directly by *BS 7671*. Therefore, where installed equipment is not covered by an appropriate British or Harmonised Standard as required by Regulation 133.1.1, or is used outside the scope of its Standard, the designer is required to provide confirmation that the resultant degree of safety is **not** less than that achieved by full compliance with *BS 7671* (Regulation 133.1.3 refers). Generally this will require supporting information to be provided by the manufacturer of the equipment, such as copies of the manufacturer's type testing documents or results of any third party testing. As a result, the presence of the departure should be indicated on the form as required by Regulation 120.3 and full details of the departure and the relevant supporting evidence should be provided on a separate numbered page(s). and appended to the certificate. Where no departures have been sanctioned, enter 'None'.

A departure must not result in a lesser degree of safety than that obtained by compliance with *BS 7671* (Regulation 120.3, 133.1.3 and 133.5 refer).

Permitted exception applied (411.3.3)

For an installation other than a dwelling, Regulation 411.3.3 permits an exception for the provision of additional protection by an RCD having a residual operating current not exceeding 30 mA, for socket-outlets with a rating not exceeding 32 A. However, where an exception is applied in accordance with Regulation 411.3.3, it must be based on a documented risk assessment which determines that RCD protection of the socket-outlet(s) is not necessary. The risk assessment (or a copy of it) must be appended to the certificate, and confirmation of that provided by inserting a 'tick' and adding the particular page numbers. It should be noted that for the purposes of *BS 7671* the risk assessment should involve an appropriately skilled person (electrically) Appendix 2, item 11, of

BS 7671 refers. Where this exception is applied the installation must be no less safe than where RCD protection is provided.

Note: the requirements of Regulation 411.3.3 are not applicable for FELV systems or reduced low voltage systems.

Verification for selectivity

Selectivity, previously known as 'discrimination', is the ability of a protective device to operate in preference to another protective device in series ensuring that the device nearest the fault is disconnected.

Therefore, where selectivity is required in accordance with Section 536.4 of *BS 7671*, details of the particular method used to verify that selectivity has been achieved should be provided on a separate numbered page(s). To confirm that the details have been appended, a 'tick' should be inserted in the space allocated and the page number(s) recorded.

Reviewed by Qualified Supervisor

Irrespective of which option for declaration is used to certify the work, for an NICEIC or ELECSA certificate the results of the inspection and testing must be reviewed and confirmed by a registered Qualified Supervisor for the contractor issuing the certificate. By signing the appropriate section, the Qualified Supervisor is confirming that the certificate has been completed satisfactorily prior to issue.

For the uncontrolled (green) version of the certificate, the completed schedules of inspection and testing should, preferably, be reviewed by another skilled person competent to verify that the recorded results are consistent with electrical installation work conforming to the requirements of *BS 7671: 2018*, as amended (Regulation 644.5 refers). The review should be confirmed by signing in the appropriate space.

2.4.2 Separate declarations for design, construction and inspection and testing (EIC only)

The following information relates to the separate declarations and so applies only to the EIC (refer to Fig 2.4.2).

Where responsibility for design, construction and inspection and testing of the electrical work is divided between the NICEIC Approved Contractor and one or more other bodies, the division of responsibility should have been established and agreed before commencement of the work.

NICEIC considers that the absence of certification for the construction, or the inspection and testing elements of the work would render the certificate invalid. If the design section of the certificate has not been completed, the notes on tshe reverse of the certificate advise the recipient to question why those responsible for the design have not certified that this important element of the work is in accordance with *BS 7671*.

Part 4: Declaration for the electrical installation work

PART 4 : DECLARATION FOR THE ELECTRICAL INSTALLATION WORK *(this option may be used where the design, construction, inspection & testing have been the responsibility of one person)*

DESIGN, CONSTRUCTION, INSPECTION & TESTING *(The extent of liability of the signatories is limited to the work detailed in PART 2)*

I, being the person responsible for the design, construction, inspection and testing of the electrical installation, particulars of which are described in PART 2, having exercised reasonable skill and care when carrying out the design and additionally where this certificate applies to an addition or alteration, having confirmed that the safety of the existing installation is not impaired, hereby CERTIFY that the design, construction, inspection and testing for which I have been responsible is to the best of my knowledge and belief in accordance with *BS 7671: 2018*, amended to .... (date) except for the departures, if any, detailed on attached page(s) (.........) (Regulations 120.3, 133.1.3 and 133.5).

• Permitted exception applied (411.3.3): Yes/NA   Risk assessment attached: (.........)   Page No(s) (.........)   • Where selectivity is required, details of the verification appended (536.4): (.........) Page No(s) (.........)

Name (capitals): ....   Signature: ....   Date: ....

REVIEWED BY QUALIFIED SUPERVISOR

Name (capitals): ....   Signature: ....   Date: ....

Fig 2.4.1 Single signature declaration for design, construction and inspection and testing

PART 4 : DECLARATION FOR THE ELECTRICAL INSTALLATION WORK *(this option may be used where the design, construction, inspection & testing have been the responsibility of one person)*

DESIGN, CONSTRUCTION, INSPECTION & TESTING *(The extent of liability of the signatories is limited to the work detailed in PART 2)*

I, being the person responsible for the design, construction, inspection and testing of the electrical installation, particulars of which are described in PART 2, having exercised reasonable skill and care when carrying out the design and additionally where this certificate applies to an addition or alteration, having confirmed that the safety of the existing installation is not impaired, hereby CERTIFY that the design, construction, inspection and testing for which I have been responsible is to the best of my knowledge and belief in accordance with *BS 7671: 2018*, amended to .... (date) except for the departures, if any, detailed on attached page(s) (.........) (Regulations 120.3, 133.1.3 and 133.5).

• Permitted exception applied (411.3.3): Yes/NA   Risk assessment attached: (.........)   Page No(s) (.........)   • Where selectivity is required, details of the verification appended (536.4): (.........) Page No(s) (.........)

Name (capitals): ....   Signature: ....   Date: ....

REVIEWED BY QUALIFIED SUPERVISOR

Name (capitals): ....   Signature: ....   Date: ....

CONSTRUCTION *(The extent of liability of the signatory is limited to the work detailed in PART 2)*

I, being the person responsible for the construction of the electrical installation, particulars of which are described in PART 2, having exercised reasonable skill and care when carrying out the construction, hereby CERTIFY that the said work for which I have been responsible is, to the best of my knowledge and belief, in accordance with *BS 7671: 2018*, amended to .... (date) except for the departures, if any, detailed on attached page(s) (.........) (Regulations 120.3 and 133.5).

Name (capitals): ....   Signature: ....   Date: ....

INSPECTION & TESTING *(The extent of liability of the signatories is limited to the work detailed in PART 2)*

I, being the person responsible for the inspection and testing of the electrical installation, particulars of which are described in PART 2, having exercised reasonable skill and care when carrying out the inspection and testing, hereby CERTIFY that the said work for which I have been responsible is, to the best of my knowledge and belief, in accordance with *BS 7671: 2018*, amended to .... (date) except for the departures, if any, detailed on attached page(s) (.........) (Regulations 120.3 and 133.5).

Name (capitals): ....   Signature: ....   Date: ....

Fig 2.4.2 Separate signature declarations for design, construction and inspection and testing

Design

This section should be completed by person(s) authorised to act on behalf of the company or companies responsible for the design element of the installation. The names and signatures of the designer(s) should be entered in the spaces provided. Where two bodies have mutual responsibility for the design, the division of responsibility should be established and agreed before work commences, and both bodies responsible must certify that the design meets the requirements of *BS 7671: 2018*, as amended.

The signatory (or signatories) to the design is certifying that, with the exception of any stated departures, the design fully meets the requirements of the current issue of *BS 7671*. If the NICEIC Approved Contractor responsible for the construction of the installation is not also responsible for its design and/or inspection and testing, it will be necessary for the contractor to liaise with the other parties responsible for those aspects in order to complete all the details required on the certificate.

As stated earlier in the guidance covering the 'Single declaration', where a permitted exception is employed in accordance with Regulation 411.3.3 and / or selectivity is provided in accordance with *BS 7671*, the designer should confirm that the relevant details are provided on an additional page(s) and appended to the certificate.

In signing this section, the signatory or signatories are accepting responsibility for the safety of the design.

Construction

The construction section should be completed by the person responsible for installing the work, which in the case of an NICEIC Electrical Installation Certificate is the Approved Contractor responsible for the construction work.

In signing this section, the signatory is certifying that the construction, except for any departure(s) sanctioned by the designer, satisfies the requirements of *BS 7671*. Such certification can only be given if the signatory has a detailed knowledge of the construction of the particular installation. If the signatory personally carried out the installation work, this presents little difficulty. If, however, the signatory is the Qualified Supervisor or other person involved in a supervisory capacity only, the level of supervision of the installation work must have been sufficient to enable the certification to be provided with confidence.

For such reasons, NICEIC and ELECSA undertake regular site visits to confirm that the number of registered Qualified Supervisors is sufficient for the number of operatives employed and the range, scope and geographical spread of work being undertaken.

Inspection and testing

The person responsible for the inspection and testing of the work should certify that the inspection and testing process has been carried out in accordance with *BS 7671: 2018, as amended.* If the signatory is the person who actually conducted the whole of the inspection and testing, this should be straightforward. However, if the process was conducted by more than one person, or if a Qualified Supervisor or someone in a similar supervisory capacity certifies the inspection and testing work on behalf of the Approved Contractor, that person must be entirely satisfied that the process was carried out thoroughly and in accordance with Part 6 of *BS 7671*, and that the results are consistent with installation work complying with all the relevant requirements of that Standard.

Certification for inspection and testing provides an assurance to the recipient that the results of the inspection and testing have been compared with the relevant criteria (Regulation 641.3. refers) and that the electrical installation work is in accordance with *BS 7671*, except for any departures sanctioned by the designer(s) and recorded on the certificate.

Reviewed by Qualified Supervisor

For NICEIC and the ELECSA certificates the results of the inspection and testing must be reviewed and confirmed by a registered Qualified Supervisor for the contractor issuing the certificate. Refer to the guidance provided earlier, under this heading, for the 'Single signature declaration'.

2.5 Part 5: Comments on the existing installation

Comments are applicable only where the work carried out is an addition or alteration to an existing installation. Any defects observed on the existing installation that may give rise to danger, should be recorded (Regulation 644.1.2 refers). Where necessary, details should be provided on an additional numbered page and appended to the certificate, and the page number(s) should be inserted in the space provided and also recorded in Part 10 . Where there are no comments to be made, insert 'None'.

PART 5 : COMMENTS ON THE EXISTING INSTALLATION *(in the case of an addition or alteration see Regulation 644.1.2)*
....
....
....
....

Fig 2.5 Comments on the existing installation

Any defects which would result in a reduced level of safety in the new work (that is a level of safety less than would be afforded by compliance with *BS 7671: 2018*) must be corrected before the new work is put into service.

2.6 Part 6: Details of the organisation(s) responsible for the electrical installation (EIC only)

This section, which is only applicable to the EIC, identifies to the recipient the organisation(s) responsible for the work certified by their representative(s) in the 'Separate declarations' for design, construction and inspection and testing, detailed earlier. The organisation(s), address(es) and postcode(s) must be given, together with their NICEIC registration number and branch number, where appropriate. Where an Approved Contractor has been responsible for more than one element, the details should be inserted in both places.

DESIGN, CONSTRUCTION, INSPECTION & TESTING
Organisation: ....
Registration No*: ....
Branch No*: ....

DESIGN
DESIGNER 1
Organisation: ....
Registration No*: ....
Branch No*: ....
Address: ....
....
....
Postcode: ....
Tel No: ....

DESIGNER 2
Organisation: ....
Registration No*: ....
Branch No*: ....

CONSTRUCTION
Organisation: ....
Registration No*: ....
Branch No*: ....
Address: ....
....
....
Postcode: ....

INSPECTION & TESTING
Organisation: ....
Registration No*: ....
Branch No*: ....
Address: ....
....
....
Postcode: ....

Fig 2.6 Details of the organisation(s) responsible for the electrical installation

2.7 Part 7: Supply characteristics and earthing arrangements

For the NICEIC and the ELECSA Domestic Electrical Installation Certificate sthe section entitled 'Supply characteristics and earthing arrangements' is contained in Part 5.

System type and earthing arrangements
TN-C-S: (.........) TN-S: (.........) TT: (.........)
Other *(state)*: ....
Supply protective device
(BS (EN) ....)
Type: (..............................) Rated current: (.........) A

Fig 2.7.1 Supply characteristics and earthing arrangements

System type and earthing arrangements

The installation earthing arrangement should be identified, which for a low voltage supply will normally be one of the following types:

- **TT:** Earthing provided by the consumer's own installation earth electrode – no earthing facility is made available to the consumer by the distributor, or if such a facility is made available, it is not used.
- **TN-S:** Earthing facility provided by the distributor for the consumer's use - provision usually by means of a connection to the supply cable sheath or a separate protective conductor in the form of a split-concentric cable or overhead conductor.
- **TN-C-S:** Earthing facility provided by the distributor, connected to the incoming supply neutral to give a Protective Multiple Earth (PME) supply, where the supply neutral and protective conductors are in the form of a Combined Neutral and Earth (CNE) conductor.

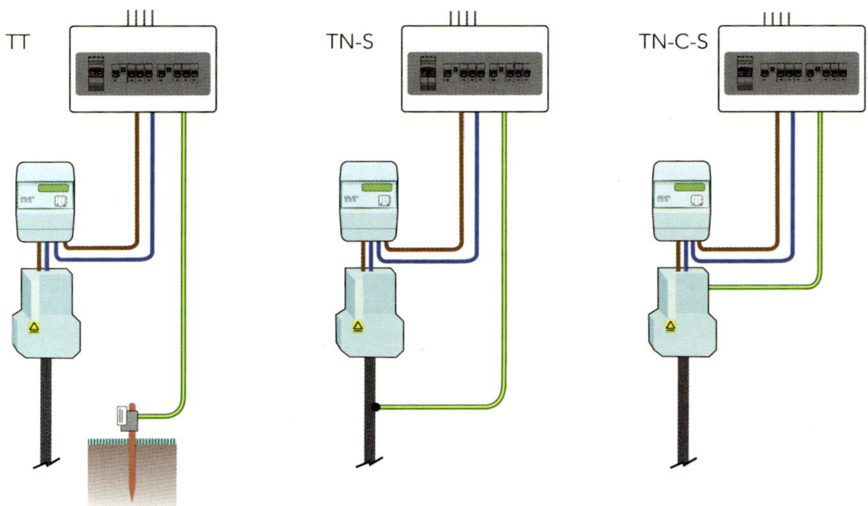

Fig 2.7.2 Common earthing arrangements

If the system is not one of the common types listed then the details of the supply system arrangement should be stated in the space provided.

Within certain types of buildings or premises, more than one system earthing arrangement may have been employed by the designer. For example, an installation in a sports complex may as a whole form part of a TN-C-S system, with the majority of the installation earthed to the PME earthing facility. For electrical design reasons, however, particular parts of the installation, such as a swimming pool area, may form part of a TT system. Similarly, the installation within a dwelling may form part of a TN-C-S system, being earthed to the PME earthing facility and another part of the installation, such as an outbuilding or garage, may be connected to a TT earthing arrangement. Where an installation has more than one earthing arrangement, each arrangement should be identified on the certificate.

Supply protective device

The information to be recorded here is the British Standard (or other appropriate Standard) product specification in terms of BS (EN) number, together with the type and the rated current (I_n).

This information should be obtained from the installation designer at an early stage, or by reference to the electrical installation records such as the certificate issued on completion of the original electrical installation. Where it has not been possible to determine the details of the supply fuse(s) by these means, confirmation of type and rating should be sought from the Distribution Network Operator (DNO).

In no circumstances should the DNO's supply fuse be withdrawn unless prior authorisation has been obtained.

Where the installation is comprised of more than one supply source, the details of the additional supply source including the supply protective device should be provided on an additional page (see 'Other sources of supply').

Number and type of live conductors			
AC	1-phase, 2-wire: (.........)		2-phase, 3-wire: (.........)
	3-phase, 3-wire: (.........)		3-phase, 4-wire: (.........)
DC	2-wire: (.........)	3-wire: (.........)	**Other:** (.....................)
Confirmation of supply polarity:			(.........)
Other sources of supply (*as detailed on attached schedule*)			Page No:(.........)

Fig 2.7.3 Number and type of live conductors

Number and type of live conductors

This section requires the nature of supply current, in terms of AC or DC and the number of the supply live conductors (including the neutral), to be identified by entering a 'tick' in the appropriate field. In most cases, an AC supply will be single-phase (two-wire) or three-phase (four-wire), but other configurations are also provided.

As the DEIC is restricted to installations having single-phase supplies up to 100 A, the options on the DEIC are reduced to 'AC, 1 phase, 2 wire'. Any supply that is not 'AC, 1 phase, 2 wire' should be recorded in the field identified as 'Other'.

Confirmation of supply polarity

Supply polarity must be confirmed, and confirmation of correct polarity should be indicated by inserting a 'tick' in the appropriate field.

Other sources of supply

Where the installation is connected to an additional source of supply, such as a standby generator or a photovoltaic source, the relevant details of the additional supply should be recorded on an additional numbered page and attached to the certificate. The page number(s) of the additional page(s) should be inserted in the space provided.

Nature of supply parameters

In this section details of the supply parameters should be recorded. For some parameters measurements are not appropriate.

Nature of supply parameters		
Nominal line voltage, U (1):	(.........) V	(1) *By enquiry, measurement, or by calculation*
Nominal line voltage to Earth, U_0 (1):	(.........) V	
Nominal frequency, f (1):	(.........) Hz	
Prospective fault current, I_{pf} (1)**:	(.........) kA	
External loop impedance, Z_e (1)**:	(.........) Ω	

Fig 2.7.4 Nature of supply parameters

Nominal voltage(s)

Measured values, should not be taken, for public two-phase and three-phase supplies in the UK, the line voltage U is 400 V and the voltage to earth U_0 is 230 V. For single-phase supplies, U is not applicable (N/A) and U_0 is 230 V.

Nominal frequency f (Hz)

In the UK the nominal low voltage supply frequency is recorded as 50 Hz, measurements should not be taken. Where doubt exists enquiry should be made to the supplier.

Prospective fault current I_{pf} (kA)

The maximum fault current likely to occur in the installation, the higher of the prospective short-circuit current and the prospective earth fault current, is the value on which the design and selection of equipment have been based. The magnitude of the prospective fault current may be obtained by 'enquiry', calculation or by 'measurement'. Detailed guidance on the measurement of prospective fault current is given in Chapter 8 'Testing'. Alternatively, the designer may contact the electricity distributor to obtain a value of prospective fault current by enquiry. Usually, neither the designer nor the person responsible for inspection and testing will have sufficient information to enable a realistic value to be calculated.

In some circumstances, testing to determine the prospective fault current at the origin or elsewhere within a domestic installation is unnecessary. A consumer unit complying with *BS EN 61439-3* is considered to be adequate to safely withstand a prospective fault current of up to 16 kA (Regulation 536.4.201 refers), and as a result of this conditional rating applying tests to determine prospective fault current at either the origin or for final circuits would, in many cases, be unnecessary (see Chapter 8 'Measurement of maximum prospective fault current'). For an additional supply source such as a standby generator, measurement of prospective fault current is generally unnecessary and in many cases impracticable.

The maximum fault current that the machine is capable of delivering is based on its constructional design and should be readily available from the manufacturer of the generator or from the documentation supplied with the generator. Likewise for a photovoltaic system a measurement of prospective fault current would not be appropriate or practical, especially as the maximum output current and short-circuit current is limited by design.

External earth fault loop impedance Z_e (Ω)

As with the prospective fault current, guidance relating to the measurement of the external earth fault loop impedance (Z_e) is given in Chapter 8. Although Z_e may be determined by enquiry, a measurement is generally still necessary in order to verify that the intended means of earthing is present and of the expected value.

Where the installation is supplied by more than one source, such as the public supply and a standby generator, the higher or highest value of external earth fault loop impedance, Z_e, should be recorded. Such information is required to verify that automatic disconnection will be achieved in accordance with *BS 7671* when the installation is being supplied by the standby generator. Information relating to additional sources may be obtained from the manufacturer's literature supplied with the additional supply source.

For safety reasons, live testing should be avoided unless it is the only practicable way to obtain the information required. For example, while a test may need to be performed to confirm that a generator is suitably connected to Earth, the test is not relevant if the source has no connection to Earth, which is often the case for a PV source. Safety relating to live testing is covered in Chapter 8.

2.8 Part 8: Particulars of installation referred to in this certificate

For the NICEIC or the ELECSA Domestic Electrical Installation Certificate the section entitled 'Particulars of installation referred to in this certificate' is contained in Part 7.

Maximum demand (load):	(.........) kVA / A *(delete as appropriate)*
Means of Earthing	
Distributor's facility:	(.........)
Installation earth electrode:	(.........)
Where an earth electrode is used insert	
Type – rod(s), tape, etc:	(..........................)
Location:	(..)
Electrode resistance to Earth:	(.........) Ω

Fig 2.8.1 Particulars of installation – Means of Earthing

Maximum demand (load)

This is not, as is sometimes assumed, the rating of the electricity distributor's cut-out fuse(s). The maximum demand is a value, expressed in amperes per phase, evaluated on the basis of the connected load (in amperes) with an allowance for diversity, and if required converted to kVA. The installation designer would normally be expected to provide the maximum demand value.

In the case of additions and alterations, it is essential that an assessment of the maximum demand of the existing installation is made prior to carrying out modifications. To this assessment of the existing maximum demand must be added all additional loads resulting from the proposed new work, taking account of diversity as appropriate, so that the suitability of the incoming electricity service to supply the proposed increase in load can be ascertained.

Means of Earthing

The particular means of earthing should be identified, whether the connection is made via the distributor's facility and/or an installation earth electrode.

Where an earth electrode is used

If the installation has an earth electrode, the type of electrode used should be inserted. Earth electrodes recognised by *BS 7671* are listed in Regulation 542.2.2.

Where a lead sheath or other metal covering of a cable is used as an earth electrode it is subject to the requirements of Regulation 542.2.5. Metal pipework forming part of a gas, flammable liquids, water or any other service utility must not be used as an earth electrode (Regulation 542.2.6 refers).

Note: Where two or more protective devices are in series, the exposed-conductive-parts may be connected to separate earth electrodes corresponding to each protective device (Regulation 411.5.1 refers).

Sufficient details, such as a drawing reference, should be given under 'location' so that skilled persons unfamiliar with the installation and building layout will be able to locate the electrode for periodic inspection and testing purposes. The earth electrode resistance to Earth must be measured and the value (in ohms) recorded in the space provided for the purpose. Guidance on the methods used to measure earth electrode resistance is given in Chapter 8 'Testing'.

Every TT system must have an installation earth electrode. Where the system is not TT and there is no earth electrode, write 'NONE' for the type and 'N/A' (Not Applicable) in the other fields. Do not leave any fields empty.

Main protective conductors

The conductor material and the cross-sectional area of the earthing conductor and the protective bonding conductors need to be recorded and confirmation of continuity provided. By inserting a 'tick' the inspector is confirming that continuity has been verified and the connections have been checked and are satisfactory. (Guidance on testing protective conductors is given in Chapter 8.)

Main protective conductors	
Earthing conductor:	
(material .... csa .... mm^2)	
Connection / continuity verified:	(.......)
Main protective bonding conductors:	
(material .... csa .... mm^2)	
Connection / continuity verified:	(.......)

Fig 2.8.2 Particulars of installation – main protective conductors

Main protective bonding connections

The main protective bonding protective bonding connections to extraneous-conductive-parts should be identified. A 'tick' confirms the protective bonding is carried out in accordance with Regulation 544.1.2.

Main protective bonding connections	
Water installation pipes:	(.........)
Gas installation pipes:	(.........)
Structural steel:	(.........)
Oil installation pipes:	(.........)
Lightning protection:	(.........)
Other *(state)*:	

Fig 2.8.3 Particulars of installation – main protective bonding connections

Main switch / Switch-fuse / Circuit-breaker/RCD

As the information that needs to be recorded in this section will depend on the design, this is normally a straightforward process of copying down the relevant details from the installation design, having checked that the installed device conforms to that design.

The 'type' of main switching device should have the relevant British Standard or BS EN number clearly marked. If, after checking both the design information and the device, no such number or any other meaningful identification of the type of the device can be found, the lack of this essential information should be referred to the designer for clarification; do not leave blank. The 'Location' of the main switching device should also be recorded. Use of a drawing reference may be appropriate.

Whilst 'No of poles' is usually 2 for a single-phase device, and 3 or 4 for a three-phase device, a variety of switching arrangements are available so this needs to be confirmed. The voltage and current rating should be inserted and where applicable the fuse rating or setting should also be inserted. Where the main switching device is an RCD the rating referred to is the rated current, (I_n) of the RCD not the rated residual operating current, $I_{\Delta n}$, of the device.

Main switch / Switch-fuse / Circuit-breaker / RCD			
Type:	(BS (EN) ....)		
Location:	(...)		
No. of poles:	(.........)	Rating / setting of device:	(.........) A
Current rating:	(.........) A	Voltage rating:	(.........) V
Where an RCD is used as the main switch			
RCD rated residual operating current, $I_{\Delta n}$:			(.........) mA
Measured operating time: (.........) ms		Rated time delay:	(.........) ms

Fig 2.8.4 Main switch / Switch fuse / Circuit-breaker / RCD

The remaining items only apply where an RCD is used as the main switch.

RCD residual operating current ($I_{\Delta n}$)

This should be determined from the design information and checked with the marking on the device. In order to minimise the risk of unwanted tripping an assessment of selectivity should have been carried out.

Measured RCD operating time (ms)

Record the measured operating time in milliseconds when tested at any current greater than or equal to the rated residual operating current, $I_{\Delta n}$. (Guidance on this test, which requires the use of an RCD test instrument, is given in Chapter 8).

Rated time delay

Where an RCD is installed as the main circuit-breaker, to provide selectivity between that device and any RCD positioned downstream of it requires an RCD classified as 'type S' (for selectivity) which incorporates a time-delay. Where this is the case, the rated time-delay in milliseconds of the RCD should be recorded.

It should be noted that the disconnection time should comply with the relevant requirements of Chapter 41.

As stated earlier, in the guidance for the 'Declaration', where it is necessary to incorporate 'Selectivity' into the electrical design the details of the verification should be provided on an additional numbered page and attached to the certificate.

2.9 Part 10: Schedules and additional pages

For the NICEIC and the ELECSA Domestic Electrical Installation Certificate the section entitled 'Schedules and additional pages' is contained in Part 8.

All schedules and additional pages appended to the certificate should be identified in the relevant sections under this heading. As required by Regulation 644.3, schedules of inspections and tests are supplied as part of NICEIC and ELECSA certificates (pages 4, 5 and 6 in the case of the EIC) and these schedules must be appended when the certificate is issued.

PART 10 : SCHEDULES AND ADDITIONAL PAGES	
Schedule of Inspections Page No(s): (....4 & 5....)	**Schedule of Circuit Details and Test Results for the installation** Page No(s): (....6....)

Additional pages, including data sheets for additional sources Page No(s): (..........)	**Special installations or locations** ***(indicated in item 11 above)*** Page No(s): (..........)	**Continuation sheets** Page No(s): (..........)

Fig 2.9 Schedules and additional pages

Where one or more of the special installations or locations contained in Part 7 of *BS 7671* form part of the work to be verified, the details, including any additional inspection items that require confirmation should be provided on an additional numbered page(s) and appended to the certificate. The page numbers of all continuation sheets appended to the certificate should be identified in the space provided.

Guidance on how to complete the schedules is given in:

- Chapter 5 'The Schedule of Circuit Details and Test Results', and
- Chapter 6 'The Schedule of Items Inspected'.

CHAPTER 3
THE MINOR ELECTRICAL INSTALLATION WORKS CERTIFICATE

- The NICEIC red Minor Electrical Installation Works Certificate (MEIWC) is available for use by NICEIC Approved Contractors.
- The NICEIC red Domestic MEIWC is available for use by NICEIC Domestic Installers.
- The ELECSA blue MEIWC is available for use by ELECSA registered contractors.

The green (uncontrolled) MEIWC is available for use by electrical contractors not registered with NICEIC or ELECSA, in particular those preparing for registration, and for those contractors registered with NICEIC or ELECSA who carry out work outside the scope of their registration.

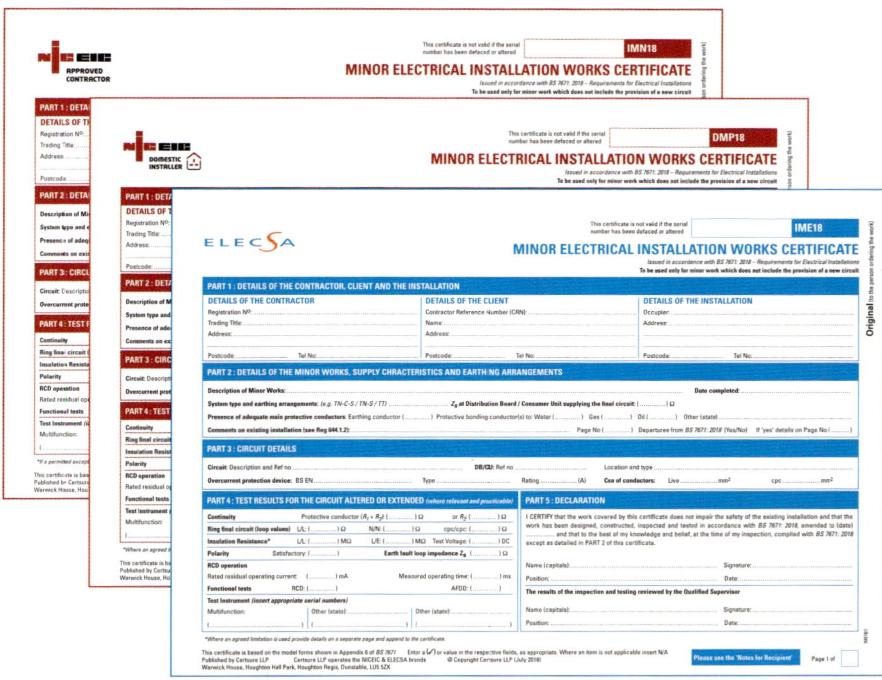

Fig (i) NICEIC and ELECSA Minor Electrical Installation Works Certificates

General

A Minor Electrical Installation Works Certificate (MEIWC) may be issued as an alternative to an Electrical Installation Certificate (EIC) or Domestic Electrical Installation Certificate (DEIC) to certify an addition or alteration to an existing installation which does not include the provision of a new circuit (Regulation 644.4.201 refers). A separate MEIWC should be issued for each circuit that has been altered or extended.

The application of the Minor Works Certificate is limited so care must be exercised to ensure that it is **not** used to certify work outside its scope. Examples of work that are within the scope of the minor works certificate include:

- The addition of a socket-outlet to an existing ring or radial final circuit.
- Work carried out on an existing lighting final circuit, including an alteration to the switching arrangements and the addition of a lighting point.
- The replacement of an accessory or luminaire associated with a single circuit*.

A single MEIWC may be used to certify more than one modification to the same circuit, but a single MEIWC must not be used to certify modifications to more than one circuit.

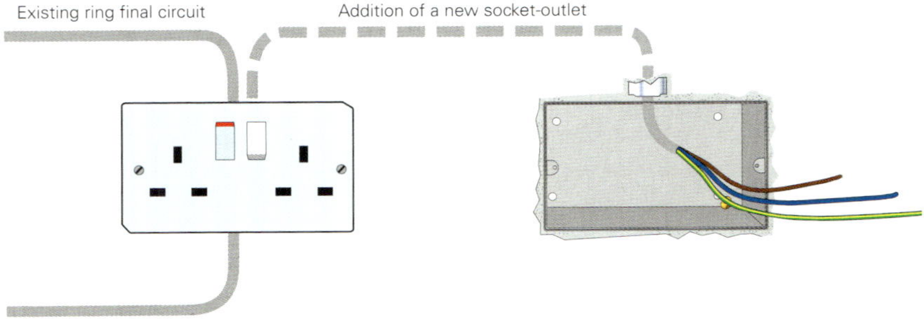

Fig (ii) Addition of a socket-outlet to an existing ring final circuit

* If preferred by a client such as a local authority, the client's own documentation (such as a combined works order and certificate) may be used instead of a Minor Works Certificate for the replacement of an accessory or luminaire, provided that the safety declaration, installation details, and the inspection and test results required to be recorded on the alternative documentation are no less comprehensive than those required to be recorded on the Minor Works Certificate. A copy of the completed documentation should be retained by the contractor.

The MEIWC may also be used to certify the replacement of a protective device in a like-for-like manner for a single circuit and, where appropriate, it may also be used to certify the replacement of a single protective device where the characteristics of the replacement device are different from the existing device. For example, the MEIWC may be used to certify the replacement of a circuit-breaker for an RCBO, or the replacement of a type B circuit-breaker with a type C circuit-breaker, as long as the skilled person completing the certificate can verify that the work (replacement) satisfies the relevant requirements of *BS 7671*, and does not impair the safety of the existing installation (Regulation 644.5 refers).

It should be noted, however, that the replacement of an item of switchgear containing more than one protective device such as a two way consumer unit is not minor works, and as such an EIC (or where appropriate a DEIC) must be issued to certify such a replacement.

Examples of acceptable and non-acceptable uses of the MEIWC are given in Appendix C of this Guide.

Before commencing the minor electrical installation work, it is necessary to check that the rating and condition of any existing equipment, including that of the electricity distributor (which may have to carry additional load), are adequate for the altered circumstances, and the earthing and protective bonding arrangements, if necessary for the protective measures applied for the safety of the work, are also adequate (Regulation 132.16 refers).

Where the existence of a dangerous or potentially dangerous situation is discovered in the existing installation (such as the absence of any necessary protective bonding where the method of fault protection is ADS), the addition or alteration should not proceed and the client should be advised immediately of the situation, preferably in writing, to satisfy the duties imposed by the *Electricity at Work Regulations 1989*. This may be achieved by the issue of an Electrical Danger Notification form to the client.

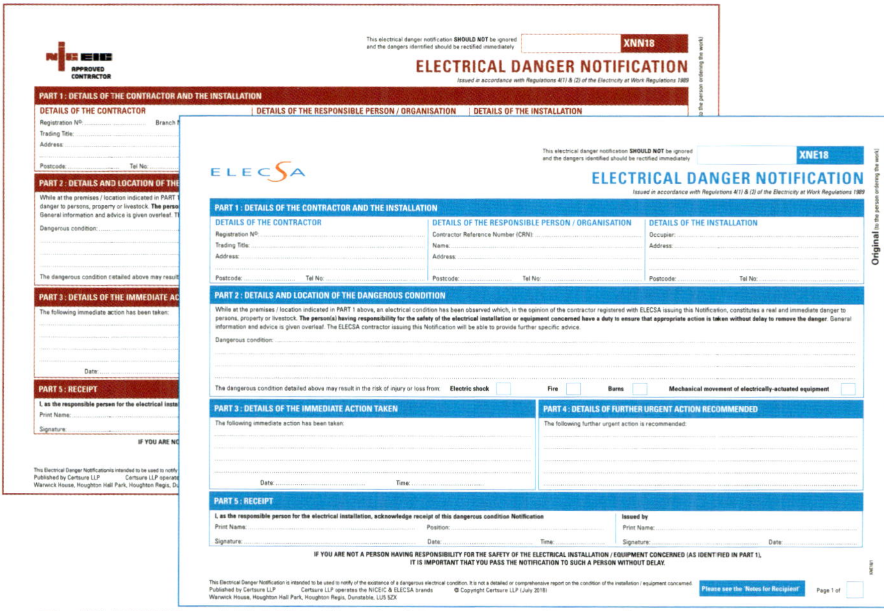
This electrical danger notification **SHOULD NOT** be ignored and the dangers identified should be rectified immediately

XNN18

ELECTRICAL DANGER NOTIFICATION

This electrical danger notification **SHOULD NOT** be ignored and the dangers identified should be rectified immediately

XNE18

ELECTRICAL DANGER NOTIFICATION

Issued in accordance with Regulations 4(1) & (2) of the Electricity at Work Regulations 1989

Original (to the person ordering the work)

PART 1 : DETAILS OF THE CONTRACTOR AND THE INSTALLATION

DETAILS OF THE CONTRACTOR	DETAILS OF THE RESPONSIBLE PERSON / ORGANISATION	DETAILS OF THE INSTALLATION
Registration Nº:	Contractor Reference Number (CRN):	Occupier:
Trading Title:	Name:	Address:
Address:	Address:	
Postcode: Tel No:	Postcode: Tel No:	Postcode: Tel No:

PART 2 : DETAILS AND LOCATION OF THE DANGEROUS CONDITION

While at the premises / location indicated in PART 1 above, an electrical condition has been observed which, in the opinion of the contractor registered with ELECSA issuing this Notification, constitutes a real and immediate danger to persons, property or livestock. **The person(s) having responsibility for the safety of the electrical installation or equipment concerned have a duty to ensure that appropriate action is taken without delay to remove the danger.** General information and advice is given overleaf. The ELECSA contractor issuing this Notification will be able to provide further specific advice.

Dangerous condition:

The dangerous condition detailed above may result in the risk of injury or loss from: **Electric shock** ☐ **Fire** ☐ **Burns** ☐ **Mechanical movement of electrically-actuated equipment** ☐

PART 3 : DETAILS OF THE IMMEDIATE ACTION TAKEN	PART 4 : DETAILS OF FURTHER URGENT ACTION RECOMMENDED
The following immediate action has been taken:	The following further urgent action is recommended:
Date: Time:	

PART 5 : RECEIPT

I, as the responsible person for the electrical installation, acknowledge receipt of this dangerous condition Notification	Issued by
Print Name: Position:	Print Name:
Signature: Date: Time:	Signature: Date:

IF YOU ARE NOT A PERSON HAVING RESPONSIBILITY FOR THE SAFETY OF THE ELECTRICAL INSTALLATION / EQUIPMENT CONCERNED (AS IDENTIFIED IN PART 1), IT IS IMPORTANT THAT YOU PASS THE NOTIFICATION TO SUCH A PERSON WITHOUT DELAY.

This Electrical Danger Notification is intended to be used to notify of the existence of a dangerous electrical condition. It is not a detailed or comprehensive report on the condition of the installation / equipment concerned.
Published by Certsure LLP Certsure LLP operates the NICEIC & ELECSA brands © Copyright Certsure LLP (July 2018)
Warwick House, Houghton Hall Park, Houghton Regis, Dunstable, LU5 5ZX

Please see the 'Notes for Recipient' Page 1 of

Fig (iii) NICEIC and ELECSA Electrical Danger Notification forms

3 Compilation of the certificate

The remainder of this chapter provides guidance on completing each of the five separate parts contained in the NICEIC and the ELECSA Minor Electrical Installation Works Certificates. The parts, should be completed by inserting the necessary text, a positive indication such as a 'Tick' or a numeric value of the measured parameter. Where an item is not applicable to the particular work, 'N/A' (meaning Not Applicable) should be entered. Where, exceptionally, a test is not practicable the client should be made aware, and in agreement with the client a 'LIM' meaning 'Limitation' should be recorded.

3.1 Part 1: Details of the contractor, client and installation

Generally, the details required here are the same as that described in Chapter 2 for the EIC. As the details of the person ordering the work (client) could be different to that of the premises in which the minor works has been undertaken, separate parts are provided to record the details. Where the details are the same they should be recorded in both places. The item entitled 'Contractor Reference Number (CRN)' is provided so that the contractor can assign a particular reference to the job or client, as appropriate. Otherwise 'N/A' should be inserted in the space.

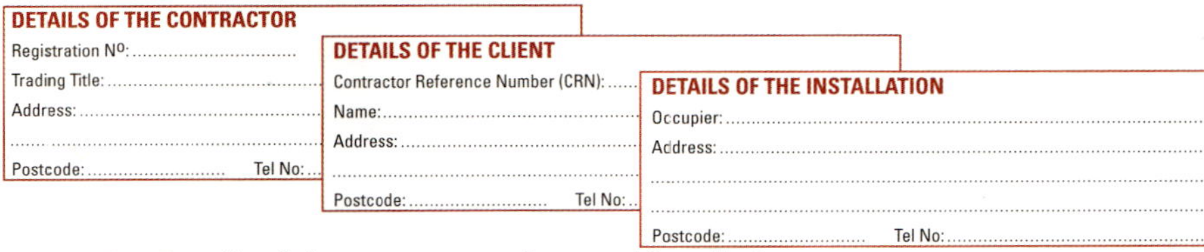

DETAILS OF THE CONTRACTOR
Registration N^{O}: ....
Trading Title: ....
Address: ....
....
Postcode: .... Tel No: ....

DETAILS OF THE CLIENT
Contractor Reference Number (CRN): ....
Name: ....
Address: ....
....
Postcode: .... Tel No: ....

DETAILS OF THE INSTALLATION
Occupier: ....
Address: ....
....
....
Postcode: .... Tel No: ....

Fig 3.1 Details of the contractor, client and installation

3.2 Part 2: Details of the minor works, supply characteristics and earthing arrangements

Description of Minor Works

A clear and accurate description of the minor works should be given together with the date the work was completed. Failure to clearly and accurately identify the work could involve the contractor in unforeseen liabilities at a later date.

Note: The NICEIC MEIWC available for use by Approved Contractors may be used to certify minor works undertaken in premises other than dwellings.

Where required a risk assessment (or a copy of it) should be appended to the minor works certificate. The risk assessment should involve electrically skilled persons (Appendix 2, item 11 of *BS 7671* refers).

System type and earthing arrangements and Z_S at the distribution board / consumer unit

Completion of these two items requires, the system type and earthing arrangements (TN-C-S / TN-S / TT) to be identified. For details on the three common types of arrangement refer to Chapter 2. In addition, the earth fault loop impedance Z_S should be determined (generally by measurement where practicable) at the distribution board or consumer unit where the circuit that is the subject of the minor works is connected. For guidance on earth fault loop impedance testing refer to Chapter 8 'Testing'.

PART 2 : DETAILS OF THE MINOR WORKS, SUPPLY CHRACTERISTICS AND EARTHING ARRANGEMENTS

Description of Minor Works*: .... **Date completed:** ....

System type and earthing arrangements: *(e.g. TN-C-S / TN-S / TT)* .... **Z_s at Distribution Board / Consumer Unit supplying the final circuit:** (........) Ω

Presence of adequate main protective conductors: Earthing conductor (........) Protective bonding conductor(s) to: Water (........) Gas (........) Oil (........) Other (*state*) ....

Comments on existing installation (see Reg 644.1.2): .... Page No (........) Departures from *BS 7671: 2018* (Yes/No) If 'yes' details on Page No (........)

Fig 3.2 Details of the minor works, supply characteristics and earthing arrangements

Presence of adequate main protective conductors

The presence and adequacy of main protective conductors should be confirmed by inspection and by continuity checks of the bonding conductors and their terminations. By inserting a 'tick' in the relevant fields the inspector is providing confirmation of the adequacy of the Earthing conductor and the adequacy of the protective equipotential bonding to the extraneous-conductive-parts identified.

The adequacy of the existing earthing and main protective bonding should be ascertained before the work is undertaken (Regulation 132.16 refers).

Comments on existing installation

Any defects observed on the existing installation whilst undertaking the minor work that may give rise to danger (non-compliances) should be recorded. Any defect that affects the safety of minor works carried out on an existing installation must be corrected before the certificate is issued (Regulation 644.1.2 refers).

Where necessary detailed comments should be provided on an additional numbered page and the page number(s) inserted in the space provided. Where there are no comments to be made, insert 'None'. It should be noted that there is no requirement to carry out a formal inspection of parts of an existing installation unrelated to the minor works, but where there is reason to believe that unrelated parts of the existing installation are in an unsatisfactory condition, you should recommend to the client that the installation (or the appropriate part of it) is inspected and tested, and an Electrical Installation Condition Report is issued.

Departures from *BS 7671: 2018*

Exceptionally, the use of new materials or inventions may lead to a departure from *BS 7671*, especially due to the development of new products not yet covered by relevant product standards.

In such circumstances, the use of a departure should be clearly identified on the certificate as required by Regulation 120.3 and full details and the justification for the departure should be provided on an additional page and attached to the certificate (Regulations 133.1.3 and 133.5 refer). Where there are no departures, as will generally be the case, insert, 'N/A'.

In signing the declaration in Part 5, the contractor is taking full responsibility for the departures and declaring that any departures identified will not result in a degree of safety less than that provided by full compliance with *BS 7671*.

3.3 Part 3: Circuit details

PART 3: CIRCUIT DETAILS			
Circuit: Description and Ref no	**DB/CU:** Ref no	Location and type	
Overcurrent protection device: BS EN .... Type	Rating .... (A)	**Csa of conductors:** Live .... mm²	cpc .... mm²

Fig 3.3 Circuit details

Circuit and distribution board / consumer unit

The information provided should enable the circuit which is the subject of the minor works to be easily located. A description and reference number for the circuit should be inserted: for example, 'Circuit No 2 – Kitchen ring circuit'. Similarly, information should be provided to enable the distribution board / consumer unit to be easily located. The level of detail required will depend on the type of installation. For small installations, such as a flat or a, typical, domestic premises, where the whole installation is supplied from a single consumer unit, a consumer unit reference number would not generally be applicable and in such cases (N/A) should be recorded. Similarly, where there is unlikely to be any ambiguity as to the location of the consumer unit, then a comment such as, for example, '1 phase 8 way c/u - under the stairs' would generally be sufficient.

In contrast, a larger installation may contain several distribution boards across multiple floors or areas, so the identification details provided should include a unique distribution board reference number and accurate location details, such as a drawing location reference, along with the type of distribution board, (for example, 3 phase - 18 way). In agreement with the client, distribution boards and / or consumer units may need to be assigned unique identification references.

Overcurrent protective device

The details of the circuit protective device should be inserted. The British Standard (or BS EN or similar) number; the type (if any) and the rated current (I_n) of the existing protective device should be recorded. For example: *BS EN 60898*; type B; 32 A.

In addition, the cross-sectional area of the altered or extended circuit conductors should be recorded in the space provided. The 'live' conductors include both the line and neutral conductors.

3.4 Part 4: Test results for the circuit altered or extended

In this part of the certificate information associated with the testing of the altered or extended circuit and the related parts of the existing installation on which the circuit depends for its safety, should be recorded.

The relevant provisions of Part 6 (Inspection and testing) of *BS 7671* must be applied in full to electrical installation work covered by a Minor Electrical Installation Works Certificate. The skilled person signing the certificate is responsible for assessing each particular situation, and for deciding whether or not it is reasonable to omit any of the tests prescribed in Part 4. Where a particular inspection or test is not applicable, this should be indicated by inserting 'N/A', meaning 'Not Applicable'. For example, completion of 'loop values' is not applicable for a radial circuit, so where the circuit is a radial 'N/A' should be inserted in those fields.

Exceptionally, where a test is not practicable, the entry should be 'LIM' meaning 'Limitation', indicating that the particular circumstances prevented the test from being carried out (see 'Insulation resistance').

PART 4 : TEST RESULTS FOR THE CIRCUIT ALTERED OR EXTENDED *(where relevant and practicable)*			
Continuity	Protective conductor $(R_1 + R_2)$: (............) Ω		or R_2: (............) Ω
Ring final circuit (loop values)	L/L: (............) Ω	N/N: (............) Ω	cpc/cpc: (............) Ω
Insulation Resistance**	L/L: (............) MΩ	L/E: (............) MΩ	Test Voltage: (............) DC
Polarity	Satisfactory: (............)	**Earth fault loop impedance Z_S**	(............) Ω

***Where an agreed limitation is used provide details on a separate page and append to the certificate.*

Fig 3.4.1 Test results for the circuit altered or extended

The following tests should be applied to the altered or extended circuit, as appropriate. For guidance on conducting the tests refer to Chapter 8 'Testing'.

Continuity

The continuity of the protective conductor(s) should be verified by using either the '$(R_1 + R_2)$' or the 'R_2' test methods, and the maximum measured value (in ohms) recorded in the appropriate space. N/A should be recorded against the method that is not used.

Where the circuit altered or extended is a 'Ring final circuit' the continuity of the individual loops should also be confirmed, and the measured loop values for the line, neutral and earth loops should be recorded in the spaces provided. If the circuit protective conductor for the 'Ring final circuit' is formed by a metallic conduit 'N/A' should be recorded in the space provided for the 'cpc' loop value.

Insulation resistance

Where practicable, insulation resistance testing should be carried out between all live conductors (line and neutral conductors) and between all live conductors and Earth; that is, the test should be performed with the protective conductors connected to the earthing arrangement. In each case the lowest value of measured insulation resistance should be recorded (in megohms), and the particular DC voltage at which the test was performed should be inserted.

As stated in Table 64 of *BS 7671*, for low voltage circuits, single or three-phase, insulation resistance testing should be performed at 500 V DC. However, where it is not reasonably practicable to disconnect equipment that may be damaged by the 500 V DC test voltage, such as in the case of a surge protective device incorporated into a fixed low voltage socket-outlet, the test may be performed at a reduced voltage of 250 V DC. However, where this is the case, the low voltage minimum insulation resistance of 1 MΩ remains applicable (Regulation 643.3.2 refers). Where insulation resistance testing is not carried out between live conductors it must be recorded as a 'LIM' meaning 'Limitation'.

Any limitation(s) should be agreed with the person ordering the work, before the work is undertaken. As stated in the 'footer' of the Minor Electrical Installation Works Certificate, brief details of the reasons and technical justifications for omitting a test should be provided on a separate page, and appended to the certificate.

Polarity

Once it is established that the polarity of the circuit is satisfactory a tick (✓) should be entered in the space provided. Additionally, the polarity of the incoming supply will need to be confirmed.

Earth fault loop impedance Z_S

The earth fault loop impedance (Z_S), should be determined at the furthest point on the circuit and recorded in ohms. The value of Z_S must be such as to afford automatic disconnection of the protective device within the time limit prescribed in *BS 7671*, in the event of an earth fault.

RCD operation		
Rated residual operating current: (............) mA		Measured operating time: (............) ms
Functional tests RCD: (............)		AFDD: (............)
Test Instrument ***(insert appropriate serial numbers)***		
Multifunction:	Other (state):	Other (state): ....
(...)	(...)	(...)

Fig 3.4.2 Test results for the circuit altered or extended

RCD operation

Where minor work is carried out on a circuit that is protected by an RCD (this includes an RCBO) the rated residual operating current ($I_{\Delta n}$) of the device should be inserted and the RCD should be subjected to a test according to its function and the measured operating time recorded (RCD testing is covered in Chapter 8).

An RCD installed to provide fault protection should provide disconnection within the relevant times of Chapter 41, whereas an RCD installed to provide additional protection in accordance with Regulation 415.1.1 must disconnect within 40 ms (see Regulation 643.8).

Functional tests

Irrespective of whether an RCD is installed to provide fault or additional protection a functional test, by operation of the test button, is required to confirm satisfactory operation of the device. Inserting a 'tick' (✓) confirms that a functional test has been carried out on the RCD and the device operated satisfactorily.

Where an Arc Fault Detection Device (AFDD) is connected to the circuit (subject of the minor works) to provide additional protection against fire in accordance with Regulation 532.6, a functional test should be carried out where practicable. Where the device does not incorporate a test button or where no such device is present insert 'N/A'.

Test Instruments (insert appropriate serial numbers)

The serial number of each of the instruments used to obtain the test results should be recorded. If an instrument does not have a serial number, a suitable number should be assigned and permanently marked on the instrument.

3.5 Part 5: Declaration

By completing the declaration the contractor is certifying that the work undertaken in connection with the addition or alteration does not impair the safety of the existing installation, and that the new work has been inspected and tested in accordance with *BS 7671: 2018*. In addition, the contractor is certifying that the minor works has been designed and constructed in accordance with the requirements of *BS 7671: 2018*, with the exception of any departures identified in Part 2 of the certificate.

The date of the amendment to *BS 7671: 2018* that was current at the date on which the work was carried out should be stated in the space provided in the declaration.

PART 5 : DECLARATION

I CERTIFY that the work covered by this certificate does not impair the safety of the existing installation and that the work has been designed, constructed, inspected and tested in accordance with *BS 7671: 2018*, amended to (date) .... and that to the best of my knowledge and belief, at the time of my inspection, complied with *BS 7671: 2018* except as detailed in PART 2 of this certificate.

Name (capitals): .... Signature: ....

Position: .... Date: ....

The results of the inspection and testing reviewed by the Qualified Supervisor

Name (capitals): .... Signature: ....

Position: .... Date: ....

Fig 3.5 Declaration

Name (Capitals)

The name of the skilled person providing the declaration on behalf of the contractor should be recorded in capital letters in the space provided.

Signature

The signature of the skilled person providing the declaration on behalf of the contractor, and the date of signing, should be inserted in the appropriate spaces. The signatory should be an electrically skilled person who is well informed about the design and construction of the work and competent to confirm whether the requirements of *BS 7671: 2018* have been met.

The signatory should have personally carried out and verified the work, or have undertaken detailed supervision of the work, so as to be in a position to take with confidence full responsibility for the statements made in the declaration.

Position:

The position of the signatory should be identified in the space provided (that is NICEIC or ELECSA Qualified Supervisor, director, partner, or supervising electrical engineer).

Date:

The date on which the signatory signs the declaration should be recorded in the space provided.

The results of the inspection and testing reviewed by the Qualified Supervisor

The results of the inspection, testing and assessment must be reviewed and confirmed by a registered Qualified Supervisor for the contractor issuing the certificate. By signing the appropriate section, the Qualified Supervisor is confirming that the certificate has been completed satisfactorily prior to issue (Regulation 644.5 refers).

For the uncontrolled (green) version of the certificate, the completed schedules of inspection and testing should preferably be reviewed by another skilled person, competent to verify that the recorded results are consistent with electrical installation work conforming to the requirements of *BS 7671*. The review should be confirmed by signing in the appropriate box.

CHAPTER 4
THE ELECTRICAL INSTALLATION CONDITION REPORT

General

(i) The need for periodic inspection and testing

Every electrical installation deteriorates with use and age and as such it is important for installations to be maintained in a safe and serviceable condition. For some installations, this is achieved by a system of continuous monitoring and maintenance of the installation and maintaining the appropriate records (Regulation 652.2 refers). However, for most installations this is achieved through periodic inspection undertaken by skilled person(s), competent in such work. Deficiencies recorded in the Electrical Installation Condition Report (EICR) may then be remedied such that the installation may continue to be used in safety.

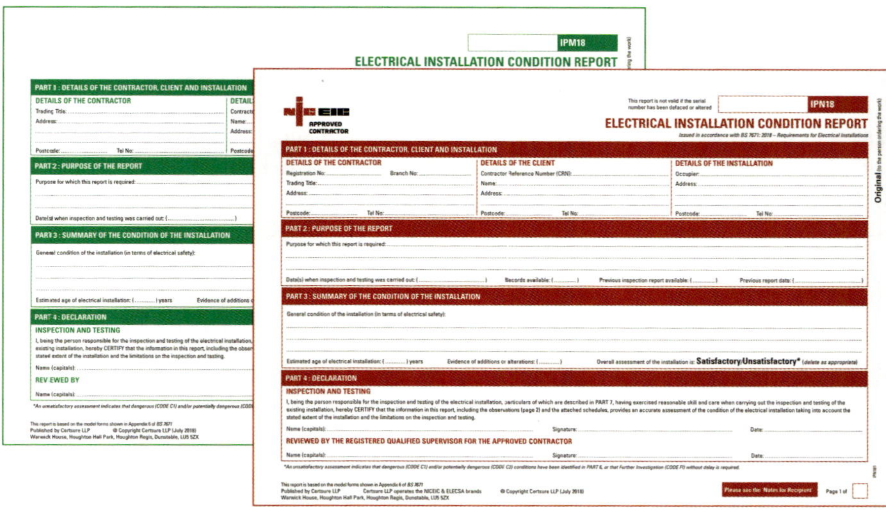

Fig (i) Electrical Installation Condition Reports

The red NICEIC Electrical Installation Condition Report is for use by NICEIC Approved Contractors, and the green (uncontrolled) Electrical Installation Condition Report form is available for use by:

- Electrical contractors not registered with NICEIC and ELECSA, in particular those preparing for registration, and
- NICEIC and ELECSA contractors or installers working outside the scope of their registration.

(ii) Purpose of periodic inspection

The main purpose of periodic inspection and testing is to detect so far as is reasonably practicable, and to report on, any factors impairing the safety of an electrical installation. The aspects to be covered, as stated in Regulation 651.2, include all of the following:

- The safety of persons and livestock against the effects of electric shock and burns
- Protection against damage to property by fire and heat arising from an installation defect
- Confirmation of the correct rating and setting of protective devices
- Confirmation of the correct rating and setting of monitoring devices
- Confirmation that the installation is not damaged or deteriorated so as to impair safety
- Identification of installation defects and non-compliances with the requirements of *BS 7671*, that may give rise to danger.

Unless the circumstances make it unavoidable (for example, if an installer has ceased trading prior to certifying an installation), an Electrical Installation Condition Report should not be issued by one contractor as a substitute for an Electrical Installation Certificate or a Domestic Electrical Installation Certificate for work carried out by another contractor.

Exceptionally, if a contractor is called upon to issue an Electrical Installation Condition Report for work designed and installed by others, the contractor should make the purpose and details of the Report absolutely clear, in order to avoid unwittingly assuming responsibility for aspects of the work of which the contractor had no knowledge and over which the contractor had no control. As stated on the Report, it would be impracticable to assess the condition of cables and conduits concealed within the fabric of the building.

(iii) Safety

Attention is drawn to the need for all persons undertaking inspection and testing to be skilled persons competent in such work (Regulation 651.5 refers).

Persons undertaking periodic inspection not only need to follow safe working procedures but must also have extensive knowledge and experience of electrical installation matters to enable them to safely and accurately assess the condition of an existing electrical installation, especially when they do not have access to the design information relating to that installation (such as diagrams and circuit details). Where no previous electrical documentation is available for the installation, such as the initial electrical certificate or previous reports, the installation will need to be investigated to determine whether it is safe for the periodic inspection and test to proceed.

If during the course of inspection or testing, a real and immediate danger is found to be present in an installation (from an exposed live part, for example), the danger should, wherever practicable, be made safe on discovery. It is not sufficient simply to draw attention to the danger when submitting a Report, it should be made safe as required by 'The Electricity at Work Regulations 1989'. In agreement with the person responsible for the safety of the installation, the appropriate action should be taken to remove the source of danger (for example, by switching off and isolating the affected part of the installation until remedied), before continuing with the inspection or testing.

(iv) Requirements

BS 7671 requires that an Electrical Installation Condition Report is issued to the person ordering the work following the in-service periodic inspection and testing of an electrical installation (Regulations 653.1 and 653.6 refer). A separate report is required for each distinct installation examined.

The Report should include the details of the parts of the installation inspected and tested and the details of any limitations (including the reasons for these limitations) imposed, together with a record of the inspections and the results of the testing (Regulation 653.2 refers).

Completion of the Report provides a formal declaration that, within the stated and agreed limitations, the details recorded, including the observations and recommendations and the completed schedules of inspection and test results, give an accurate assessment of the condition of the electrical installation at the time it was inspected and tested. For contractors not authorised to issue the NICEIC Electrical Installation Condition Report forms, similar (green) report forms based on the model form given in *BS 7671* are available from NICEIC and ELECSA.

The Electrical Installation Condition Report is, as its title indicates, a report and not a certificate. It relates to an assessment of the in-service condition of an electrical installation against the requirements of the issue of *BS 7671* current at the time of the inspection, irrespective of the age of the installation. The criteria for assessing the compliance of each part of an older installation with *BS 7671* are, therefore, the same as for new installation work. The Report is for the benefit of the person ordering the work, and of persons subsequently involved in additional or remedial work, or further inspections. The following statement contained in the introduction to *BS 7671: 2018* should be noted:

> *'Existing installations that have been installed in accordance with earlier editions of the Regulations may not comply with this edition in every respect. This does not necessarily mean that they are unsafe for continued use or require upgrading.'*

An EICR is to be used only for reporting on the condition of an existing installation. It must not be used instead of an Electrical Installation Certificate or a Domestic Electrical Installation Certificate for certifying new electrical installation work, or as a substitute for a Minor Electrical Installation Works Certificate for certifying an addition or an alteration to an existing installation and must not be used to certify the replacement of a distribution board or consumer unit.

For installations having more than one distribution board, or more circuits than can be recorded on Page 6, one or more additional pages of the 'Schedule of Circuit Details and Test Results' will be required (see Chapter 5). Continuation sheets are available from NICEIC and ELECSA.

(v) Periodic inspection and testing procedures

The procedures for periodic inspection and testing differ in some respects from those for the initial verification of new installation work. This is because the subject of a periodic inspection is usually an installation which has been energised and in use for some time. Particular attention therefore needs to be given during the inspection process to assessing the condition of the installation in respect of damage, deterioration, external and other such factors that may influence the safety of the installation.

In view of this, the range and order of tests is likely to differ for periodic testing to that which is given in *BS 7671* for new installation work (including work involving an addition or an alteration). For comparison purposes, Fig (ii) shows both the sequence for initial testing and that recommended for periodic testing.

Testing is covered in Chapter 8 of this Guide.

Initial sequence of tests	Periodic tests
For initial testing from *BS 7671*	**For periodic testing**
(i) Before the supply is connected, or with the supply disconnected as appropriate • **Continuity of conductors** Including - main and supplementary - equipotential bonding - conductors of ring final circuits. • **Insulation resistance** Including - low voltage circuits - FELV, SELV or PELV or by separation of circuits - Resistance / impedance of floors and walls. • **Polarity** • **Earth electrode resistance (where practicable)** (ii) With the electrical supply connected (re-check polarity before further testing). • **Protection by automatic disconnection of the supply** Including - Earth fault loop impedance. - Prospective fault current. - RCD testing. • **Verification of phase sequence.** • **Functional testing** Including - RCDs, AFDDs and functional testing of switchgear and controlgear, assemblies, drives, controls and interlocks. • **Verification of voltage drop (if required)**	With the supply disconnected as appropriate • Continuity of circuit protective conductors. • Continuity of main and supplementary equipotential bonding conductors. • Continuity of ring final circuit conductors. • Insulation resistance. • Polarity. • Earth electrode resistance. • Earth fault loop impedance. • Verification of phase sequence. • Functional test of RCDs and AFDDs. • Functional test of circuit-breakers, isolators and switching devices. • Verification of voltage drop (if necessary). Note 1: The person carrying out the testing is required to decide which of the above tests are appropriate by using their experience and knowledge of the installation being inspected and tested and by consulting any available records. Note 2: The tests need not be carried out in the order shown.

Note: For initial verification, the tests performed before connecting the supply (or with the supply disconnected) should be completed in the order shown (Regulation 643.1 refers).

Fig (ii) The general sequence of tests for initial verification and periodic inspection

4 Compiling the form

Parts of the Electrical Installation Condition Report form are identical or similar to parts of the Electrical Installation Certificate and the Domestic Electrical Installation Certificate, the preparation of which is described in Chapter 2. Where each of these parts is reached in this chapter, the reader is referred to Chapter 2 in order to avoid repetition.

Generally, irrespective of the method of compilation (by hand, computer software or using NICEIC's online certification system) the Report must be completed by inserting the necessary text, or a 'tick' to indicate that the task has been completed and that the result was satisfactory, or a numeric value of the measured parameter, or by entering 'N/A' meaning 'Not Applicable', where appropriate.

If a limitation on a particular inspection or test has been agreed, and has been recorded in Part 7, the appropriate field should be completed by inserting 'LIM', indicating that an agreed limitation has prevented the inspection or test being carried out. The total number of pages which make up the Report must be inserted in the box provided at the foot of each of the pages on the right-hand side.

4.1 Part 1: Details of the contractor, client and installation

The details required are the same as those described in Chapter 2. The item entitled 'Contractor Reference Number (CRN)' is provided so that the contractor can assign a particular reference to the job or client, as appropriate. Otherwise 'N/A' should be inserted in the space.

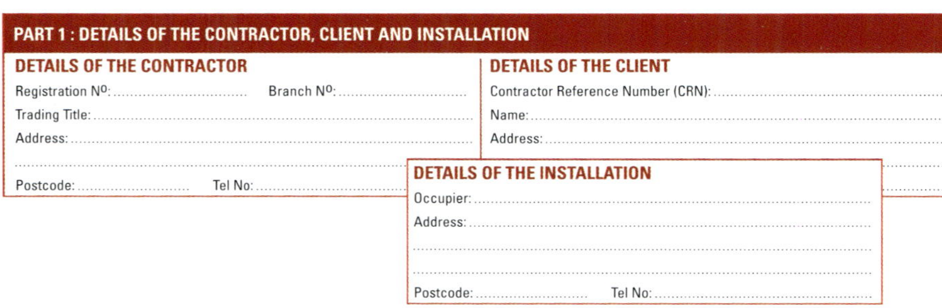
PART 1 : DETAILS OF THE CONTRACTOR, CLIENT AND INSTALLATION

DETAILS OF THE CONTRACTOR
Registration Nº: Branch Nº:
Trading Title:
Address:
Postcode: Tel No:

DETAILS OF THE CLIENT
Contractor Reference Number (CRN):
Name:
Address:

DETAILS OF THE INSTALLATION
Occupier:
Address:
Postcode: Tel No:

Fig 4.1 Details of the contractor, client and installation

4.2 Part 2: Purpose of the Report

It is important, for all concerned, that a clear statement of the purpose of the Report is made in this part of the form. There may be a range of reasons why a report is required, for example, in connection with a property purchase or as part of an application for an entertainments licence. Alternatively, a report on the condition of an installation may be required following a fire or flood, or because the period recommended in the original EIC or DEIC, or a previous report, has expired.

In addition, the date when the inspection and testing was undertaken (or dates if the inspection and testing were undertaken over a number of days) should be inserted. This is intended to be the date(s) when the inspections and testing was actually carried out, rather than when the Report was issued.

PART 2 : PURPOSE OF THE REPORT

Purpose for which this report is required: ....

....

....

Date(s) when inspection and testing was carried out: (......................................) Records available: (............)

....

....

....

Previous inspection report available: (............) Previous report date: (......................................)

Fig 4.2 Purpose of the Report

As required by Regulation 651.1, those undertaking the inspection should take account of any previous electrical documentation, such as the original electrical certificate (EIC or DEIC) or any subsequent inspection reports. Reference to the previous Report may reveal, for example, items previously coded as unsatisfactory that have not been addressed. Furthermore, having access to the results from the original electrical certificate and/or any subsequent report may help determine the rate at which any deterioration has occurred to the installation.

Where records are available a 'tick' should be inserted in the space provided. Similarly, if the previous inspection Report is available this should be identified and the date of the previous Report recorded. As stated earlier, if no documentation is available, the installation should be investigated to determine whether the inspection can proceed safely (Regulation 651.1 refers).

4.3 Part 3: Summary of the condition of the installation

The summary should adequately describe the overall condition of the installation, taking into account the specific observations made. It is essential to provide a clear summary of the condition of the installation having considered, for example:

- The adequacy of the earthing and bonding
- The type(s) of wiring system, and its condition
- The extent of any wear and tear, damage or other deterioration

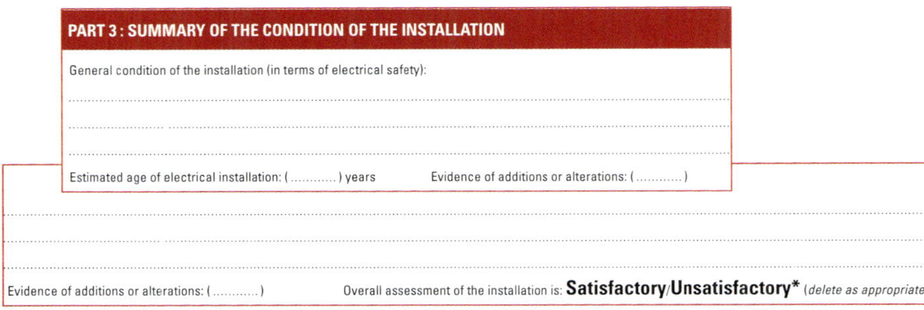
PART 3 : SUMMARY OF THE CONDITION OF THE INSTALLATION

General condition of the installation (in terms of electrical safety):

Estimated age of electrical installation: (............) years Evidence of additions or alterations: (............)

Evidence of additions or alterations: (............) Overall assessment of the installation is: **Satisfactory/Unsatisfactory*** *(delete as appropriate)*

Fig 4.3 Summary of the condition of the installation

Minimal descriptions such as 'poor', and superficial statements such as 'Recommend a rewire' are considered unacceptable as they do not indicate the true condition of the installation. In some circumstances, it may be necessary or appropriate to explain the implications of an Electrical Installation Condition Report in a covering letter, for the benefit of recipients who require additional advice and guidance about their installation. For example, where an installation has deteriorated or been damaged to such an extent that its safe serviceable life can reasonably be considered to be at an end, a recommendation for renewal should be made in a covering letter, giving adequate supporting reasons. Reference to the covering letter should be made in the Report.

Estimated age of electrical installation: This may be determined from information relating to the original installation. Otherwise a reasonable estimate should be made by other means, such as the colours used to identify the wiring or the appearance of the installed equipment.

Evidence of additions or alterations: If there is clear evidence as to whether or not additions or alterations have been made to the installation since the previous inspection (or from new, if no subsequent information is available), this should be indicated by inserting 'Yes' or 'No', as appropriate.

After due consideration, the overall condition of the installation should be identified as either 'Satisfactory' or 'Unsatisfactory' by deleting one term as appropriate.

As indicated by the asterisk, which relates to the text in the 'footer' of the page, if any observation in the Report is given a 'code C1' or 'code C2' recommendation, or if an observation requires further investigation (FI) to determine whether danger or potential danger exists, the overall assessment of the installation should be identified as being 'Unsatisfactory'.

Where the space provided is insufficient for describing the condition of the installation, the summary should be continued on an additional page(s) and the page number(s) should be recorded here and in Part 11 'Schedules and additional pages'.

4.4 Part 4: Declaration

The inspector is required to provide confirmation that the Report gives an accurate assessment of the condition of the installation within the installation details and limitations specified in Part 7. The declaration should be signed by the skilled person competent to undertake the inspection, and then by a skilled person, competent to review the Report and verify if it has been correctly compiled (Regulation 653.5 refers).

For an NICEIC Electrical Installation Condition Report, the reviewer must be a registered Qualified Supervisor for the Approved Contractor. Where the Qualified Supervisor has personally conducted the inspection and testing, and has compiled the Report, that person should sign in both places.

For the uncontrolled (green) version of the EICR, the completed Report should preferably be reviewed by another skilled person or persons, competent to confirm that the declared overall condition of the electrical installation is consistent with the inspection and testing results, and with the observations and recommendations for action (if any) made in the Report.

PART 4 : DECLARATION

INSPECTION AND TESTING

I, being the person responsible for the inspection and testing of the electrical installation, particulars of which are described in PART 7, having exercised reasonable skill and care when carrying out the inspection and testing of the existing installation, hereby CERTIFY that the information in this report, including the observations (page 2) and the attached schedules, provides an accurate assessment of the condition of the electrical installation taking into account the stated extent of the installation and the limitations on the inspection and testing.

Name (capitals): .... Signature: .... Date: ....

REVIEWED BY THE REGISTERED QUALIFIED SUPERVISOR FOR THE APPROVED CONTRACTOR

Name (capitals): .... Signature: .... Date: ....

Fig 4.4 Declaration for the inspection and testing of the electrical installation

PART 5 : NEXT INSPECTION

I/We (as indicated on page 1) recommend, subject to the necessary remedial work being taken, this installation should be further inspected and tested after an interval of not more than .... years/months* (delete as appropriate)

Give reason for recommendation: ....

Fig 4.5 Next inspection

4.5 Part 5: Next inspection

In accordance with Regulation 653.4, an appropriate recommendation for the time interval to the next inspection should be inserted. The interval period should take into account any legislative or licensing requirements, as well as the level of maintenance the installation is likely to receive.

Justification of the interval recommended should be provided by a brief supporting statement. For example, if a condition report is required 5-yearly as part of the licence agreement for a House of Multiple Occupancy (HMO), then the recommendation should reflect this period and the supporting statement should identify this as the reason for the recommended interval to the next inspection. A shorter period may be considered appropriate if an installation is old or perhaps due to a change of usage is now subject to more rapid deterioration than it has been previously. In all cases, the inspector must provide a statement of justification for the recommended interval.

Guidance on the maximum interval recommended between initial verification and the first periodic inspection for a range of installations is given in Appendix D of this Guide. The information given should be used as a guide only and the intervals stated should, where necessary, be adjusted to meet the particular circumstances.

The recommendation for the interval to the next inspection is conditional upon all observations that have been attributed a Classification code C1 (Danger present) being remedied immediately and all observations that have been attributed a code C2 (Potentially dangerous) or FI (Further investigation required) being remedied or investigated respectively as a matter of urgency.

4.6 Part 6: Observations and recommendations for actions to be taken

Any items inspected or test results that have been identified as being dangerous (code C1) or potentially dangerous (code C2), or that would enhance safety if improved (code C3), or that require further investigation without delay to determine whether danger or potential danger exists (code FI), should be brought forward to this section of the Report in the form of observations on the safety of the installation.

Table 4.1 Classification codes

Classification codes	Meaning
C1	Danger present
C2	Potentially dangerous
C3	Improvement recommended
FI	Further investigation required

The observations and classifications should take due account of the results of the inspection and testing, and be based on the requirements of the issue of *BS 7671* current at the time of the inspection (not the requirements of an earlier standard current at the time the installation was originally constructed).

The top of the section should be completed either with a 'tick' to indicate that 'there are no items adversely affecting electrical safety' or 'N/A', should be inserted (meaning 'Not Applicable') and the observations and recommendations for action should then be recorded in a concise and easily-understandable manner.

Referring to the Schedule of Items Inspected (see PART 10), the attached Schedule of Circuit Details and Test Results (see PART 12), and subject to any agreed limitations listed in PART 7:
There are no items adversely affecting electrical safety (.......), OR The following observations and recommendations for action are made:

Fig 4.6.1 Observations and recommendations for actions to be taken

Each recorded observation should describe a specific defect or omission in the electrical installation. The observation should detail what the situation is, and not what is considered necessary to put it right, for example:

'Excessive mechanical damage to the enclosure of distribution board F2/7' recommend replacement' would be considered appropriate. Whereas simply stating 'distribution board F2/7 needs replacing' without providing any technical justification, would not.

It should be noted that an EICR is a factual report on the condition of an electrical installation, not a proposal for remedial work. Only observations that can be supported by one or more regulations in the current issue of *BS 7671* should be recorded. The particular regulation number(s) need not be entered on the Report (unless specifically required by the client), but should serve to remind the inspector that it is only compliance with *BS 7671* that is to be considered. Observations based solely on personal preference or 'custom and practice' should not be included.

PART 6 : OBSERVATIONS AND RECOMMENDATIONS FOR ACTIONS TO BE TAKEN

CODES: *One of the following Codes, as appropriate, has been allocated to each of the observations made below to indicate to the person(s) responsible for the electrical installation the degree of urgency for remedial action*	CODE C1 'Danger Present' Risk of injury. Immediate remedial action required	CODE C2 'Potentially Dangerous' Urgent remedial action required	CODE C3 'Improvement Recommended'	CODE FI 'Further Investigation Required Without Delay'

Referring to the Schedule of Items Inspected (see PART 10), the attached Schedule of Circuit Details and Test Results (see PART 12), and subject to any agreed limitations listed in PART 7:

There are no items adversely affecting electrical safety (.......), OR The following observations and recommendations for action are made:

Item No	Observation(s)	Code	Location Reference
(.......)	(..........)	(.........)	(..........)
(.......)	(..........)	(.........)	(..........)
(.......)	(..........)	(.........)	(..........)
(.......)	(..........)	(.........)	(..........)
(.......)	(..........)	(.........)	(..........)
(.......)	(..........)	(.........)	(..........)
(.......)	(..........)	(.........)	(..........)
(.......)	(..........)	(.........)	(..........)
(.......)	(..........)	(.........)	(..........)
(.......)	(..........)	(.........)	(..........)
(.......)	(..........)	(.........)	(..........)
(.......)	(..........)	(.........)	(..........)
(.......)	(..........)	(.........)	(..........)
(.......)	(..........)	(.........)	(..........)
(.......)	(..........)	(.........)	(..........)
(.......)	(..........)	(.........)	(..........)
(.......)	(..........)	(.........)	(..........)
(.......)	(..........)	(.........)	(..........)

Additional pages? (.............) State page numbers: (..........)

'Immediate action required' for items: (..........) **'Improvement recommended'** for items: (..........)

'Urgent remedial action required' for items: (..........) **'Further investigation required without delay'** for items: (..........)

Fig 4.6.2 Observations and recommendations for actions to be taken

Observations and classifications should be itemised and given a numerical reference under the left-hand column headed 'Item No'. Each observation must be attributed with a Classification code C1, C2, C3 or FI, which is to be recorded in the appropriate column 'Code'. Additionally, the column 'Location Reference' is provided so that items can, where appropriate, be assigned a particular drawing location reference or distribution board and circuit reference. The reference assigned should enable the item to be located easily, especially by those tasked with undertaking the remedial works at a later date.

It should usually be possible for the inspector to attribute an outcome to each observation without requiring further investigation. Further investigation required (FI) should not be called for unless the inspector has reason to believe that further investigation of an observation is likely to reveal danger or potential danger that could not be fully identified at the time.

If there is reasonable doubt as to whether danger or potential danger exists, the condition of the installation must be reported to be 'Unsatisfactory'.

The person ordering the Report should be advised that any observations for which further investigation is considered necessary should be investigated without delay after the inspection, by a skilled person or persons, competent in such work.

If there are more observations than can be recorded in the space provided, an additional numbered page(s) should be provided as necessary. For NICEIC Electrical Installation Condition Reports, additional pages should be identified by the unique Electrical Installation Condition Report serial number. Continuation sheets for this purpose are available from NICEIC.

Each observation must be given an appropriate Classification code, C1, C2, C3 and FI. One, and only one, of the standard Classification codes should be attributed to each observation. If more than one classification could be applied to an observation, only the most serious classification should be used (Code C1 being the most serious). The Classification codes should be used as follows:

Classification codes:

Appendix G of this Guide provides examples of the application of Classification codes.

- **Code C1 (Danger present):** to be used to indicate that real and immediate danger exists, requiring immediate remedial action. The persons using the installation are at immediate risk. The person ordering the Report should be advised to take action without delay to remedy the observed deficiency in the installation, or to take other appropriate action (such as switching off and isolating the affected parts of the installation) to remove the danger. The inspector should ensure the danger is, wherever practicable, made safe on discovery and not wait for the full Report to be issued before giving this advice. An NICEIC 'Electrical Danger Notification' form may be issued to provide the person ordering the work with a formal notification of the safety issue and the remedial work required.

- **Code C2 (Potentially dangerous):** to be used to indicate that, whilst an observed deficiency is not considered to be dangerous at the time of the periodic inspection, it would become a real and immediate danger if a fault or other foreseeable event was to occur in the installation or connected equipment, and that urgent remedial action is required. The person ordering the Report should be advised that, whilst the safety of those using the installation may not be at immediate risk, remedial action should be taken as a matter of urgency, by skilled person(s) competent in electrical installation work, to remove the source of potential danger.

- **Code C3 (Improvement recommended):** to be used to indicate that, whilst an observed deficiency is not considered to be a source of immediate or potential danger, improvement would contribute to enhancing the safety of the electrical installation. For example, an older installation designed to an earlier edition of *BS 7671* may not have been provided with additional protection by an RCD having a residual current rating not exceeding 30 mA for cables concealed in walls. Whilst the installation may remain safe for continued use, the safety of the installation would be enhanced if additional protection were provided.

- **Code FI (Further investigation required):** to be used to indicate that an observation requires further investigation to determine whether danger or potential danger exists. It should usually be possible for the inspector to attribute an outcome to each observation without requiring further investigation. Further investigation required (FI) should not be called for unless the inspector has reason to believe that further investigation of an observation is likely to reveal danger or potential danger that could not be fully identified at the time.

Note: Non-compliances with the requirements of the current edition of *BS 7671* that do not give rise to danger or need improvement do not need to be recorded on the Report (*Electrical Safety First, Best Practice Guide 4* refers).

If there is reasonable doubt as to whether danger or potential danger exists, the condition of the installation must be reported to be 'Unsatisfactory'.

The person ordering the Report should be advised that any observations for which further investigation is considered necessary should be investigated without delay after the inspection, by a skilled person or persons, competent in such work.

It is entirely a matter for the skilled person conducting the inspection to decide on the Classification code to be attributed to a given observation. The skilled person's judgement should not be unduly influenced by the person ordering the work. The persons signing the Report are fully responsible for its content and accuracy.

For the benefit of the recipient an explanation of each Classification code is given on the reverse of NICEIC Electrical Installation Condition Report forms (and the equivalent green forms).

Where no items adversely affecting safety have been observed and recorded, 'None' should be entered in each of the four fields at the bottom of the section. Where one or more items that do adversely affect safety have been observed and recorded, the fields should be completed as follows:

- Observations that have been attributed a Classification code C1 (Danger present) should be highlighted by inserting the relevant item numbers in the first field adjacent to the words 'Immediate remedial work required for items:'
- Observations that have been attributed a Classification code C2 (Potentially dangerous) should be highlighted by inserting the relevant item numbers in the second field adjacent to the words 'Urgent remedial action required for items:'
- Observations that have been attributed a Classification code C3 (Improvement recommended) should be highlighted by inserting the relevant item numbers in the fourth field adjacent to the words 'Improvement recommended for items:'
- Observations for which further investigation code FI is required should be highlighted by inserting the relevant item numbers in the third field adjacent to the words 'Further investigation required for items:'

Further guidance on the application of the Classification codes can be found in the Electrical Safety First's Best Practice Guide 4 - *Electrical installation condition reporting: Classification codes for domestic and similar electrical installations* - produced in collaboration with other leading industry bodies including NICEIC and ELECSA.

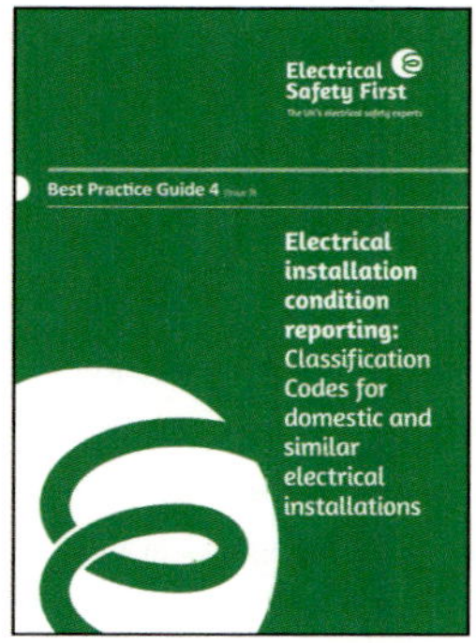

Fig 4.6.3 Best Practice Guide 4 – Electrical Installation condition reporting

The Guide can be downloaded from: **www.electricalsafetyfirst.org.uk**

Although the guidance in the Best Practice Guide is for 'domestic and similar electrical installations', the same principles can be applied to other types of electrical installation, such as in some commercial and industrial premises.

4.7 Part 7: Details and limitations of the inspection and testing

Before commencing an inspection, it is essential to agree with the client the exact extent of the installation to be inspected, taking into account the requirements of any third party involved (such as a licensing authority). It might be necessary to make a cursory inspection of the installation and available inspection records before discussing with the client the scope of the detailed inspection required, and the available options.

Details of the installation covered by this Report

The agreed details of the installation covered by the inspection should be fully recorded. Inspection and testing may apply to whole of the installation, or it may only apply to a particular part of the installation. The part could, for example, be a particular area (such as the second floor only) or particular circuits (such as lighting circuits only). In some circumstances, it may be beneficial to record parts of the electrical installation excluded from the inspection. For example, all high bay lighting or the fire alarm system. The detail will need to be tailored to accurately reflect the work undertaken. In some instances, the details may need to be continued on an additional page and appended to the Report.

Agreed limitations including the reasons, if any, on the inspection and testing

The limitations imposed on the inspection and testing should also be explained to, and agreed with, the person ordering the inspection (client) before work commences. As highlighted by the text included in Part 7, cables and wiring systems that are not accessible will not be inspected unless the client specifically requests otherwise. However, inspection should be made within accessible roof spaces housing electrical equipment which needs to be maintained, such as pumps, motors or inverters. All limitations should be agreed with the client before commencing the inspection and clearly recorded (including the reasons for these limitations) on the Report.

The name of the person(s) that agreed the limitations should be recorded in the space provided. Neither the contractor nor the person ordering the inspection should set unnecessary limitations on the work. The aim should be to produce a description that is both fair to the client and reasonable for the contractor. It should be remembered that the greater the limitations applying to a report, the less its value in terms of feedback on the safety of the installation.

Operational limitations including the reasons

Details of any circumstances that have prevented the inspection and testing from being performed as previously agreed with the person ordering the work, such as inability to gain access to a part of an installation for reasons beyond the control of the inspector, should be provided. Where necessary, details of the particular circumstances causing the operational limitation should be provided on a separate numbered page and attached to the Report.

Extent of sampling

Periodic inspection must include a thorough visual inspection of terminations at accessible electrical equipment and accessories covered by the Report. In most cases, a sampling process is applied to the detailed inspection. Depending on the size of installation, it may also be appropriate to apply a sampling process to the testing element of the work.

Wherever sampling is intended, the initial degree of sampling should be agreed with the person ordering the inspection (or their representative) before the work commences. The agreed sample for the inspection and the testing elements should be appropriate for the type and size of installation and areas of highest risk should be targeted. For example, in a building where parts of the electrical installation are considerably older than other parts, the sample size chosen for the older parts should be greater than that chosen for other parts of the installation.

Should the actual degree of sampling carried out be different from that agreed initially (because, for example, increased sampling was necessary due to deficiencies being found during the initial sampling), the actual degree should be recorded on the Report, generally as a percentage. Details of the sampling technique and identification of the equipment selected for inspection and testing on a sampling basis should also be fully identified, on an additional page if necessary.

Liabilities arising from inadequate description:
It should be appreciated that failure to clearly describe and record the details of the installation inspected and the agreed limitations (including the technical justification for these limitations) of the inspection and testing could involve the contractor in unforeseen liabilities at a later date. For example, if a contractor failed to clearly record that all high bay lighting had been excluded from a periodic inspection, the contractor might be held responsible if an electrical fault in that part of the installation later resulted in danger or damage.

If there is insufficient space to provide full details of the work, the details should be provided on an additional numbered page and attached to the Report.

4.8 Part 8: Supply characteristics and earthing arrangements

The supply characteristics and earthing arrangements should be obtained and inserted generally as described for the Electrical Installation Certificate in Chapter 2 of this Guide.

PART 8 : SUPPLY CHARACTERISTICS AND EARTHING ARRANGEMENTS

System type and earthing arrangements	**Number and type of live conductors**
TN-C-S: (.........) TN-S: (.........) TT: (.........)	**AC** 1-phase, 2-wire: (.........) 2-phase, 3-wire: (.........)
Other *(state)*:	3-phase, 3-wire: (.........) 3-phase, 4-wire: (.........)
Supply protective device	**DC** 2-wire: (.........) 3-wire: (.........) **Other:** (.....................)
(BS (EN) ....)	Confirmation of supply polarity: (.........)
Type: (..) Rated current: (.........) A	Other sources of supply (*as detailed on attached schedule*) Page No:(.........)

Nature of supply parameters		
Nominal line voltage, U (1):	(.........) V	*(1) By enquiry, measurement, or by calculation*
Nominal line voltage to Earth, U_0 (1):	(.........) V	
Nominal frequency, f (1):	(.........) Hz	
Prospective fault current, I_{pf} (1)*:	(.........) kA	
External loop impedance, Z_e (1)*:	(.........) Ω	

Fig 4.8 Supply characteristics and earthing arrangements

Wherever practicable the value for the external earth fault loop impedance Z_e should be measured in order to verify the intended means of earthing is present and of a suitable value.

However, due to operational considerations it may not be practicable to disconnect main protective bonding conductors to measure external earth fault loop impedance Z_e. For example, in premises such as a hospital or where production lines are in constant operation, it might not be possible to isolate the installation to perform a measurement of Z_e. Where this is the case, the testing should be performed with the main protective conductors connected and this should be identified as a limitation.

Alternatively, a value of Z_e may be obtained from records of the installation, provided inspection confirms the means of earthing is present and suitable for the installation. Such records should be available from the person responsible for the safety of the installation. Where this is the case, the words 'by enquiry' should be included adjacent to the value of Z_e, and the operational restriction should be recorded as part of the agreed limitations.

To avoid unnecessary live testing, wherever practicable prospective fault current should be obtained by enquiry. Guidance on prospective fault current is given Chapter 2, and guidance on 'live' testing is given in Chapter 8.

4.9 Part 9: Particulars of the installation referred to in this Report

The particulars required here should be ascertained and inserted as previously described in Chapter 2 of this Guide, and for the test methods that need be applied refer to Chapter 8.

PART 9 : PARTICULARS OF INSTALLATION REFERRED TO IN THIS REPORT

Means of Earthing	**Main protective conductors**	**Main protective bonding connections**
Distributor's facility: (.........)	Earthing conductor:	Water installation pipes: (.........)
Installation earth electrode: (.........)	(material .... csa .... mm²)	Gas installation pipes: (.........)
Where an earth electrode is used insert	Connection / continuity verified: (.......)	Structural steel: (.........)
Type – rod(s), tape, etc: (.........................)		Oil installation pipes: (.........)
Location: (...)	Main protective bonding conductors:	Lightning protection: (.........)
Electrode resistance to Earth: (.........) Ω	(material .... csa .... mm²)	Other *(state)*:
	Connection / continuity verified: (.......)	....

Main switch / Switch-fuse / Circuit-breaker / RCD			
Type:	(BS (EN) ....)		
Location:	(..)		
No. of poles:	(.........)	Rating / setting of device:	(.........) A
Current rating:	(.........) A	Voltage rating	(.........) V
Where an RCD is used as the main switch			
RCD rated residual operating current, $I_{\Delta n}$:			(.........) mA
Measured operating time: (.........) ms		Rated time delay:	(.........) ms

Fig 4.9 Particulars of the installation referred to in this Report

4.10 Part 11: Schedules and additional pages

This section is provided for the purpose of recording additional data sheets, including all 'Continuation sheets' relating to the Report. Additional pages may be required where there is a need to record supplementary information about, for example, particular aspects of the installation or its source(s), which may include data sheets relating to the characteristics of additional sources. Where the Report includes a special installation or location an additional page(s) will need to be provided to confirm the particular requirements of the relevant section of Part 7 of *BS 7671* have been met.

PART 11 : SCHEDULES AND ADDITIONAL PAGES		
Schedule of Inspections Page No(s): (4 & 5)	Schedule of Circuit Details and Test Results for the installation Page No(s): (6)	Additional pages, including data sheets for additional sources Page No(s): (..........)
Special installations or locations *(indicated in item 9. above)* Page No(s): (..........)	Continuation sheets Page No(s): (..........)	

Fig 4.10.1 Schedules and additional pages

Where additional pages are used the page numbers should be recorded in the appropriate space. (Continuation sheets are available from NICEIC and ELECSA.).

Additional pages should be given the same unique serial number as the other pages of the Report. Superseded versions of the continuation sheets must not be used in conjunction with these reports.

For the uncontrolled (green) version of the Electrical Installation Condition Report, the Contractor's Reference Number may be used to identify all additional pages and Continuation sheets.

NICEIC Domestic Electrical Installation Condition Reports are available for use by NICEIC Approved Contractors for the purpose of reporting on the condition of an existing electrical installation in a domestic property or similar. Detailed information on periodic inspection and testing of domestic premises is provided in the NICEIC and ELECSA publication: *Domestic Periodic Inspection, Testing and Reporting*.

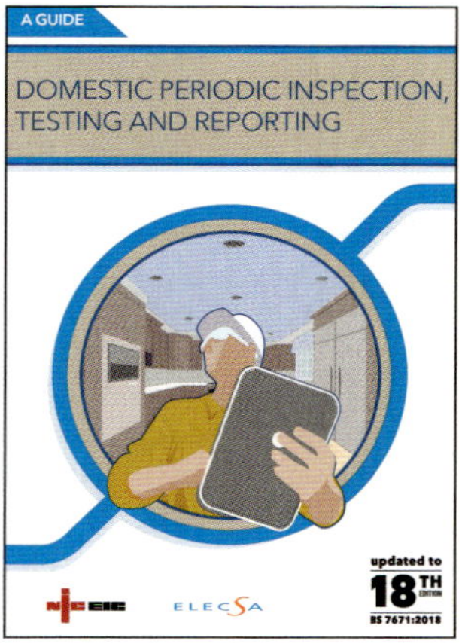

Fig 4.10.2 NICEIC and ELECSA publication:
Domestic Periodic Inspection, Testing and Reporting

CHAPTER 5
THE SCHEDULE OF CIRCUIT DETAILS AND TEST RESULTS (for new installation work and condition reporting)

The Schedule of Circuit Details and Test Results (for new installation work and condition reporting)

5.1 General

This Chapter gives guidance on how to complete the 'Schedule of Circuit Details and Test Results' which form part of NICEIC and ELECSA certificates and reports. The guidance is equally applicable if the schedules form part of an Electrical Installation Certificate (EIC) or an Electrical Installation Condition Report (EICR) and is also generally applicable to schedules that form part of the domestic versions.

Electrical certificates and reports must be accompanied by schedules of test results as required by Regulations 644.3 and 653.2 respectively. One combined 'Schedule of Circuit Details and Test Results' is provided as part of each NICEIC and ELECSA certificate or report. The first part of the schedule is used to record information about the design, such as the composition of each circuit and the type and rating of each protective device, and the second part is used to record the test results for each of those circuits. At least one schedule must be compiled for each switchboard, distribution board, consumer unit, control panel and the like within the installation covered by the EIC or EICR. Reference to the 'distribution board' in this Chapter should be taken as meaning any of the aforementioned items of switchgear, or composite assemblies of such items.

Where the installation is comprised of more circuits than can be recorded on a single schedule or where the installation has more than one distribution board, additional schedules are required in the form of continuation sheets (refer to 'Continuation sheets' in this Chapter).

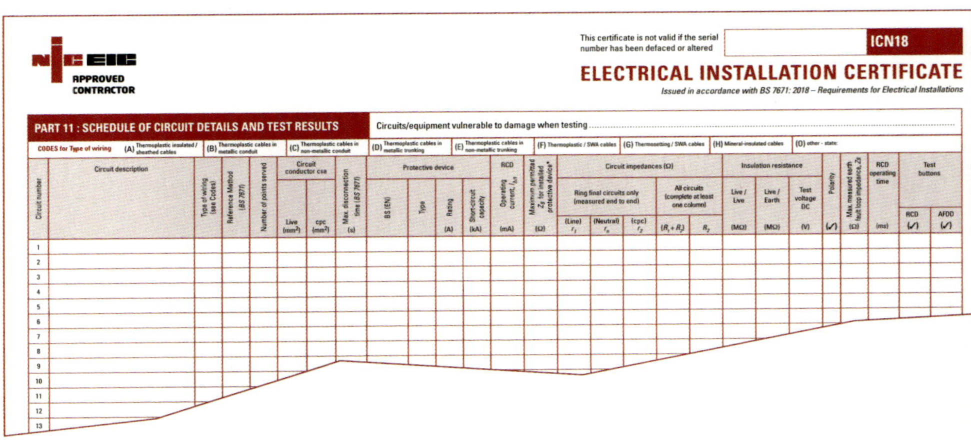
APPROVED CONTRACTOR

This certificate is not valid if the serial number has been defaced or altered | ICN18

ELECTRICAL INSTALLATION CERTIFICATE

Issued in accordance with BS 7671: 2018 – Requirements for Electrical Installations

PART 11 : SCHEDULE OF CIRCUIT DETAILS AND TEST RESULTS | Circuits/equipment vulnerable to damage when testing ....

CODES for Type of wiring (A) Thermoplastic insulated/sheathed cables (B) Thermoplastic cables in metallic conduit (C) Thermoplastic cables in non-metallic conduit (D) Thermoplastic cables in metallic trunking (E) Thermoplastic cables in non-metallic trunking (F) Thermoplastic / SWA cables (G) Thermosetting / SWA cables (H) Mineral-insulated cables (O) other - state:

Circuit number	Circuit description	Type of wiring (see Codes)	Reference Method (BS 7671)	Number of points served	Circuit conductor csa: Live (mm²)	Circuit conductor csa: cpc (mm²)	Max. disconnection time (BS 7671) (s)	Protective device: BS (EN)	Protective device: Type	Protective device: Rating (A)	Protective device: Short-circuit capacity (kA)	RCD: Operating current, $I_{\Delta n}$ (mA)	Maximum permitted Z_s for installed protective device* (Ω)	Circuit impedances (Ω): Ring final circuits only (measured end to end) (Line) r_1	(Neutral) r_n	(cpc) r_2	Circuit impedances (Ω): All circuits (complete at least one column) (R_1+R_2)	R_2	Insulation resistance: Live / Live (MΩ)	Live / Earth (MΩ)	Test voltage DC (V)	Polarity (✓)	Max. measured earth fault loop impedance, Z_s (Ω)	RCD operating time (ms)	Test buttons: RCD (✓)	Test buttons: AFDD (✓)
1																										
2																										
3																										
4																										
5																										
6																										
7																										
8																										
9																										
10																										
11																										
12																										
13																										

Fig 5.1.1 The Schedule of Circuit Details and Test Results

Circuits / equipment vulnerable to damage when testing

At the top of the schedule, space is allocated to identify any circuit(s) which has equipment connected that is vulnerable to damage from testing. For example, electronic devices such as dimmers, programmers or socket-outlets incorporating surge protective devices are likely to suffer damage if exposed to insulation resistance testing at 500 V DC. Identifying circuits that have such equipment connected is intended to support persons who have to carry out tests on the circuits in the future, such as those involved in maintenance work or periodic inspection. The relevant circuit number should be entered and where necessary, additional information should be provided on a separate page and attached to the schedule. In addition, an advisory notice should be placed at the distribution board or consumer unit (Regulation 514.9.1 refers).

Circuit description

The details of each outgoing circuit must be entered. The entry should be as descriptive as possible so that it cannot be confused with any other circuit of the distribution board. Ambiguous descriptions such as 'lighting' or 'power' would not be acceptable. The line should be identified using a scheme which is compatible with that used at the actual distribution board on site. For example: for a three-phase board the description should include a line reference such as '1L1, 1L2, 1L3'.

5.2 Circuit details

The schedule combines both the circuit details and the test results for those circuits. This section looks at the circuit details that need to be recorded.

…tion	Type of wiring (see Codes)	Reference Method (*BS 7671*)	Number of points served	Circuit conductor csa		Max. disconnection time (*BS 7671*) (s)	Protec…	
				Live (mm^2)	cpc (mm^2)		BS (EN)	Typ…

Fig 5.2.1 Circuit details

Type of wiring

Details of the type of wiring, such as 'thermoplastic insulated /sheathed' or 'thermosetting' should be identified by inserting the appropriate code. Where code 'O' (other) is used, a note must be added in the space provided in the table to identify the specific type of wiring.

CODES for Type of wiring	(A) Thermoplastic insulated / sheathed cables	(B) Thermoplastic cables in metallic conduit	(C) Thermoplastic cables in non-metallic conduit	(D) Thermoplastic cables in metallic trunking

Fig 5.2.1 Codes for Type of wiring

Reference methods

The Reference methods used for the installation of the circuit, as detailed in Appendix 4 of *BS 7671*, should be recorded here. The appropriate Reference method is selected based on the Installation method (Table 4A2 of Appendix 4 of *BS 7671* refers). Installation methods 100 – 103 inclusive, are not assigned a Reference method so the appropriate Installation method number should be recorded in the column.

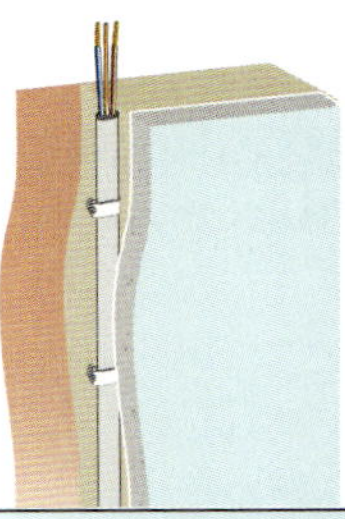

Ref A: Multi-core cables in thermally insulated wall

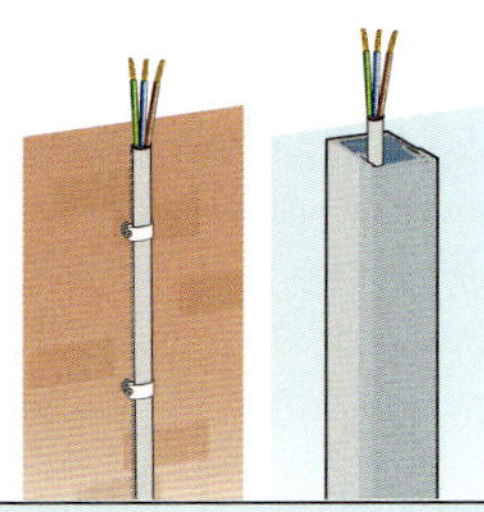

Ref B: Cables enclosed in trunking or conduit

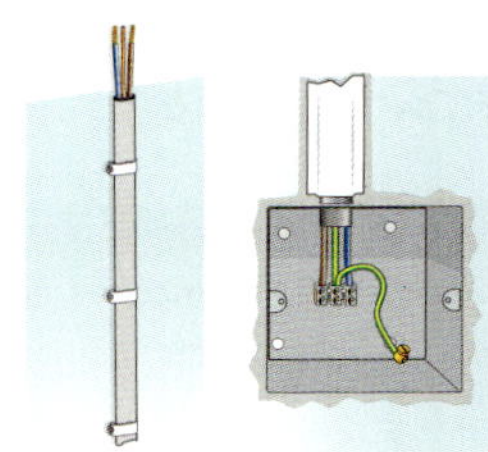

Ref C: Clipped direct

Ref D: Cables laid in the ground within a duct

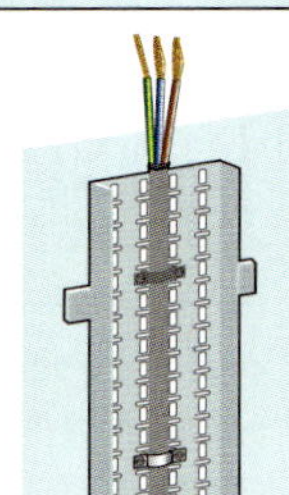

Ref E or F: On a perforated tray

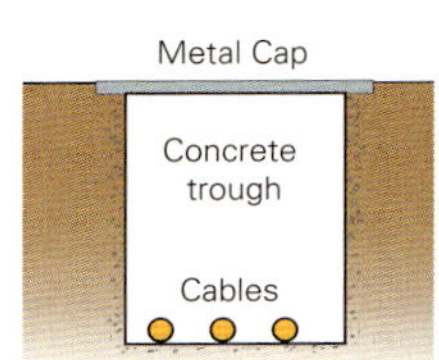

Ref E or F: Enclosed in an infloor concrete trough

Fig 5.2.2 Examples of Reference methods

Number of points served

Record the number of items of current-using equipment, socket-outlets or similar connected to the circuit. A double or multiple socket-outlet or a light fitting comprised of more than one lampholder should be recorded as a single point.

Circuit conductors (csa)

Completion of this column is straightforward for the live conductors (line and neutral), and for the circuit protective conductor (cpc) if this is a single-core cable or a core of a multi-core cable. Where the cpc is provided by the steel-wire armouring of an armoured cable, the copper sheath of a mineral insulated cable, or by steel conduit or trunking it should be indicated by its type, such as 'Arm', 'Sheath', 'Con', 'Trunk' and so on, rather than by its cross-sectional area.

In specifying such types of protective conductor, the designer of the circuit will have taken responsibility for ensuring its suitability for that purpose.

Maximum disconnection time permitted by *BS 7671*

The appropriate value from Chapter 41 of *BS 7671* should be inserted. The maximum permitted disconnection times specified in Table 41.1 are applicable to final circuits with a rated current not exceeding:

- 63 A for one or more socket-outlets, and
- 32 A supplying only fixed connected current-using equipment.

For a distribution circuit or a circuit not covered by Table 41.1, Regulation 411.3.2.3 permits a maximum disconnection time of 5 seconds for a TN system, and where such a circuit is connected to a TT system the maximum disconnection time permitted is 1 second (Regulation 411.3.2.4 refers).

Protective device

The following columns relate to the protective device, and the information required for the first four columns is generally obtained directly from the device as shown in Fig 5.2.4.

Max. disconnection time (*BS 7671*) (s)	Protective device				RCD	Maximum permitted Z_S for installed protective device*	
	BS (EN)	Type	Rating (A)	Short-circuit capacity (kA)	Operating current, $I_{\Delta n}$ (mA)	(Ω)	Rin (me (Line) r_1

Fig 5.2.3 Circuit details

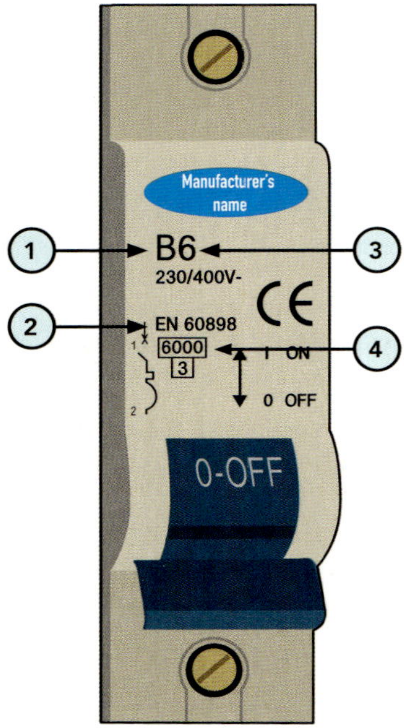

Fig 5.2.4 Protective device

1: Type:

In the case of a circuit-breaker to *BS EN 60898*, the 'type' may be 'B', 'C' or 'D'. Where the device does not have a 'type' as is the case for *BS 3036* (rewireable fuses), an entry of N/A would be appropriate.

2: BS (EN):

In this column the particular Standard that the device is designed to must be identified. Only the number needs recording. For the circuit-breaker shown, '60898' should be inserted.

3: Rating:

The rating of the device should be recorded in Amperes and as the units are indicated only the value needs inserting. For the example shown '6' should be inserted.

4: Short-circuit capacity:

The maximum fault current that the device can interrupt safely should be recorded in kA. For the device shown this is 6000 A, which is 6 kA, and as the units are in 'kA' only the '6' needs inserting.

RCD: Operating current, $I_{\Delta n}$ (mA)

The figure to be inserted in this column is the rated residual operating (tripping) current, in mA, of any RCD that may be in circuit. If there is none present, then 'N/A' should be entered.

Maximum Z_S permitted by *BS 7671* (Ω)

It should be noted that maximum values of earth fault loop impedance for protective devices can vary between manufacturers. Therefore, as stated in Appendix 3 of *BS 7671*, wherever possible designers should use the manufacturers' specific data. Alternatively, the values provided in *BS 7671* may be used.

For each protective device, the maximum value of Z_S permitted by Chapter 41 of *BS 7671* should be recorded in this column; 'corrected' values should not be inserted. Where the value is taken from a source other than *BS 7671*, such as where the designer provides specific information obtained from the manufacturer of the protective device or their literature, the particular source should be inserted in the space provided in the 'footer' of the schedule. Where necessary information such as the relevant tables from which the information was obtained should be attached on a separate page and appended to the schedule.

5.3 Test results

The remaining columns on the schedule are used to record the results of tests carried out on the circuits. The test results for each outgoing circuit must be entered in the row corresponding to the circuit. The information to be entered under each column heading is as follows. (Guidance on performing each test is described in Chapter 8.)

D current, $I_{\Delta n}$)	Maximum permitted Z_S for installed protective device*	Circuit impedances (Ω)					Insulati	
		Ring final circuits only (measured end to end)			All circuits (complete at least one column)		Live / Live	
	(Ω)	(Line) r_1	(Neutral) r_n	(cpc) r_2	$(R_1 + R_2)$	R_2	(MΩ)	

Fig 5.3.1 Test results: Circuit impedances

Circuit impedances (Ω)

The continuity of ring final conductors and circuit protective conductors should be verified by measurement, as required by Regulation 643.2.1, and the values should be recorded in the appropriate part of this section. The section contains two parts; the first part is applicable for ring final circuit(s) only, whilst the other part is applicable to all circuits.

Ring final circuits only (measured end-to-end)

Where a ring final circuit is installed the resistance of each of the circuit's conductors, the line (r_1), the neutral (r_n) and the cpc (r_2) should be measured from end-to-end and the values (in ohms) should be recorded in the three columns provided. For circuits other than ring final circuits, 'N/A' should be entered in these columns.

All circuits (complete at least one column)

An entry in at least one of the two columns ($R_1 + R_2$) or R_2, is necessary for all circuits, including ring final circuits.

($R_1 + R_2$): Enter the measured resistance, in ohms, of the line conductor and circuit protective conductor connected together as described for the measurement of ($R_1 + R_2$) in Chapter 8 (see 'continuity of protective conductors' or, for ring final circuits, 'continuity of ring final circuit conductors').

Note: For a ring final circuit, these test procedures also confirm that polarity is correct.

(R_2): Where continuity has been determined using the 'wander lead' method, or where for other reasons it is not practicable to record ($R_1 + R_2$), it is permissible to record R_2 values in this column.

If, for technically justifiable reasons, data is not recorded in either of the two columns of the schedule forming part of an Electrical Installation Condition Report, 'Lim' (meaning limitation) should be entered. In such cases, an explanation of why the testing could not be performed should be given in 'PART 7' of the Report.

Note: Such a situation should not arise in the case of new work covered by an Electrical Installation Certificate or Domestic Electrical Installation Certificate.

Insulation resistance

		Insulation resistance					
All circuits mplete at least ne column)		Live / Live	Live / Earth	Test voltage DC	Polarity	Max. measured earth fault loop impedance, Z_S	RCD operati time
R_2)	R_2	(MΩ)	(MΩ)	(V)	(✓)	(Ω)	(ms)

Fig 5.3.2 Insulation resistance

There are three columns under this heading and the information required for each is described as follows:

Live / Live

The lowest insulation resistance measured between each line conductor of the circuit and between each line and neutral conductor of the circuit should be recorded.

Live / Earth

Measurements should be taken between each live conductor (this includes the neutral conductor) and the protective conductor, ensuring that the protective conductor is connected to the earthing arrangement before performing the test. The lowest measured value should be recorded in the column. For the purposes of the test line and neutral conductors may be connected together (Regulation 643.3.1 refers).

As shown on the schedule, all measured values of insulation resistance must be recorded in megohms (MΩ). As the units are indicated on the schedule only the measured value needs inserting in the space. If the measured insulation resistance exceeds the maximum range of the instrument being used to perform the test, record the measurement as being greater than the maximum range by, for example '> 200' or '200+' or similar. Do not record the value as '∞' or 'infinity'.

Test voltage DC

The third column under 'Insulation resistance' requires the DC voltage used to perform the insulation resistance test to be recorded. The test voltage applicable is based on the nominal voltage of the circuit and is given in Table 64 of *BS 7671*. For low voltage circuits, single or three-phase, the test is required to be performed at 500 V DC, but in certain circumstances this test voltage may be reduced. For example, where it is not reasonably practicable to disconnect equipment that may be damaged by the 500 V DC insulation resistance test, such as in the case of a surge protective device incorporated into a fixed low voltage socket-outlet, the test may be performed at a reduced voltage of 250 V DC. However, in such circumstances the low voltage minimum insulation resistance of 1 MΩ remains applicable (Regulation 643.3.2 refers).

The information required in the remaining test columns on the schedule is highlighted below, and described as follows.

stance		Polarity	Max. measured earth fault loop impedance, Zs	RCD operating time	Test buttons	
	Test voltage DC (V)	(✓)	(Ω)	(ms)	RCD (✓)	AFDD (✓)

Fig 5.3.3 Test results

Polarity

The polarity of each circuit must be confirmed before the circuits are connected to the supply and energised. Correct circuit polarity is confirmed by a combination of inspection and testing and indicated by inserting a tick in this column.

It is important to appreciate that an indication of incorrect polarity must never be included on a schedule of test results for new installation work because defects revealed during testing must be corrected before the certificate is issued (Regulation 644.1.1 refers). Only in the case of an Electrical Installation Condition Report can it ever be acceptable to issue a schedule of test results showing defects such as incorrect polarity (Regulation 653.2 refers). However, even in the event of incorrect polarity being discovered during a periodic inspection it should, as a code C1, be the subject of immediate remedial action and be made safe.

In addition, to the confirmation of circuit polarity required here, the polarity of the incoming live supply must be confirmed as part of the 'Supply characteristics' before the installation is energised (or re-energised). Otherwise if the supply polarity was incorrect then all the installation circuits connected to the supply would become affected and persons and livestock would be at risk of shock.

Maximum measured earth fault loop impedance, Z_s

Before undertaking earth fault loop impedance testing, the continuity of conductors should be confirmed.

Where, as in most cases, the protective measure is automatic disconnection of supply (ADS), the highest value of earth fault loop impedance should be recorded for each circuit. This is the earth fault loop impedance measured at the point or accessory which is electrically most remote from the origin of the circuit. Where appropriate the value may be determined by an alternative method, such as by calculation (Regulation 643.7.3 refers).

Note: A table of 'Maximum earth fault loop impedance values for overcurrent protective devices in common use, for fault protection' is provided within each pad of NICEIC and ELECSA certificates and reports. However, these values are corrected values for verifying measured readings and should not be inserted in this column.

RCD operating time

Any RCD connected to the circuit must be tested to verify its effectiveness (see Regulations 643.7 and 643.8 respectively). The RCD test should be undertaken as described in Chapter 8 of this Guide and the measured value inserted. Verification of the result should be carried out according to the protection provided. For additional protection the measured operating time of the RCD should not exceed 40 ms when tested. Where the RCD is installed to provide fault protection the disconnection should be achieved within the appropriate time stated in Chapter 41 of *BS 7671*. For guidance on testing and verification of the test results refer to Chapter 8.

Test buttons

Two columns are provided to record the outcomes of functional (manual) tests carried out on RCDs and Arc Fault Detection Devices (AFDDs), where provided.

In the case of an Electrical Installation Certificate covering a new installation, or an addition or alteration, where an RCD or AFDD is present only correct operation of the integral test facility, indicated by a tick (✓) is acceptable. However, in the case of a periodic inspection of an existing installation, either a tick (✓) or a cross (✗) could be entered, as appropriate.

Where the schedule is completed as part of an EICR and the test results identify circuit defects, the defects should be carried forward to PART 6 of the Report in the form of observations on the safety of the installation and coded appropriately. Guidance on condition reporting is contained in Chapter 4.

Details of the distribution board and inspector

DISTRIBUTION BOARD (DB) DETAILS	DB designation:..
(to be completed in every case)	Location of DB:..

Fig 5.3.4 Details of the distribution board

Distribution board (DB) details:
The distribution board (DB) details required here must be completed in every case:

DB designation:
The designation may be a reference number or other identifying mark and should be unique to the particular board.

Location of DB:
The location of the distribution board (or consumer unit) should be clearly identified. Where appropriate a drawing location reference may be inserted.

Tested by

TESTED BY	Name (capitals):	Position:........................
	Signature:...	Date:............................

Fig 5.3.5 Tested by

Name (capitals):
The name of the skilled person who carried out the testing on behalf of the contractor should be recorded in capital letters, in the space provided.

Signature:
By signing, the skilled person is accepting responsibility for the details recorded on the schedule.

Position:
The position of the signatory should be identified in the space provided (that is NICEIC or ELECSA Qualified Supervisor, director, partner, or supervising electrical engineer).

Date:
The date on which the testing was carried out should be recorded in the space provided.

To be completed only if the DB is not connected directly to the origin of the installation
Where the distribution board (or consumer unit) is connected directly to the origin the details recorded under the 'Supply characteristics' will provide all the necessary information about the supply to the distribution board. Where this is the case N/A should be written across this section.

However, where the distribution board (or consumer unit) is not connected directly to the origin, the following details must be provided.

TO BE COMPLETED ONLY IF THE DB IS NOT CONNECTED DIRECTLY TO THE ORIGIN OF THE INSTALLATION			
Supply to DB is from: (...)		Nominal voltage: (.........) V	No. of Phases: (.........)
Overcurrent protection device for the distribution circuit	Type: (BS EN ....)	Rating: (.........) A	

Fig 5.3.6 Additional details required when the DB is not connected directly to the origin

In the case of a domestic installation, the details of the distribution circuit supplying a consumer unit located remotely from the origin should be recorded on the first row of the schedule as indicated by the 'asterisk'. In addition, where necessary, the prospective fault current should be determined at the remote consumer unit and recorded in the space provided under the schedule.

It should be noted that for a domestic (household) or similar premises, the value of prospective fault current I_{pf} would not need to be determined where a consumer unit to *BS EN 61439-3* is installed and the maximum prospective fault current declared by the Distribution Network Operator is 16 kA (Appendix 14 of *BS 7671* refers). Where this is the case 16 kA should be inserted.

For a schedule forming part of an EIC or EICR the following additional details are required for a distribution board or consumer unit located remotely.
Supply to DB is from:
Details of where the distribution circuit feeding the distribution board (DB) is supplied from should be recorded; for example, the particular DB reference, such as DB 4 (ground floor). This designation should be different from that of any other distribution board forming part of the installation.

The 'Nominal voltage' and 'No. of phases' are straightforward. For a low voltage AC distribution circuit, it will typically be '3' for a 400 V three-phase supply and '1' for a 230 V single-phase supply.

Overcurrent protective device for the distribution circuit

The type and rating of the overcurrent device protecting the distribution circuit that feeds the DB should be inserted: for example; type *BS (EN)*: '*BS 88-2*', Rating: '63 A'.

Associated RCD (if any)

Where the supply is protected by an RCD; the product specification number should be inserted, for example '*61008*', the number of poles ('2', '3', or '4') and the residual rated operating (tripping) current of the device ($I_{\Delta n}$) in units of mA.

An RCD test should be performed as described in Chapter 8 of this Guide and the operating time of the RCD should be inserted.

Associated RCD (if any)	Type: (BS EN ....)	No. of poles: (.........)	$I_{\Delta n}$ (.........) mA	Operating time (.........) ms
Characteristics at this DB	Confirmation of supply polarity: (.........)	Phase sequence confirmed (where appropriate): (.........)	Z_s (.........) Ω	I_{pf} (.........) kA

Fig 5.3.7 Additional details required when the DB is not connected directly to the origin

Where an RCD is installed to protect a distribution circuit the relevant requirements of Section 536 of *BS 7671* in relation to 'Selectivity' will need to be satisfied. A circuit that supplies a distribution board or consumer unit should be installed so that the need to provide additional protection by an RCD having a residual current rating of 30 mA is avoided. For these reasons, steel-wire armoured cables are commonly used to supply remote distribution boards.

Characteristics at this distribution board

At the remote distribution board (or consumer unit), the supply polarity and, where appropriate, the supply phase sequence should be confirmed. At the incoming supply terminals of the distribution board or control panel, the earth fault loop impedance (Z_s), commonly referred to as Z_{db} should be measured or determined by an alternative method such as calculation (Regulation 643.7.3 refers). Similarly, the maximum prospective fault current (I_{pf}) which is the greater of either the prospective value of short-circuit current or the prospective value of earth fault current should, where necessary, be measured (or determined by an alternative method) and the value recorded. Guidance on the tests and verifying the results is given in Chapter 8

Test instruments

In this section the serial number of each of the instruments used to obtain the test results recorded on the schedule (or continuation sheet) should be inserted. If any instrument does not have a serial number, a suitable number should be assigned and permanently marked on the instrument. To ensure that test instruments are safe for use and provide reliable results, it is essential that contractors have systems in place to monitor the on-going accuracy of their test instruments (guidance on test instrument safety is given in Chapter 8 'Testing').

TEST INSTRUMENTS (enter serial number against each instrument used)	
Multi-function:	Continuity:
(..)	(..)
Insulation resistance:	Earth fault loop impedance:
(..)	(..)
Earth electrode resistance:	RCD:
(..)	(..)

Fig 5.3.8 Test instruments

5.4 Continuation sheets

One combined schedule of circuit details and test results is supplied as part of each NICEIC and ELECSA electrical certificate and each Report (and the equivalent green forms).

Where the installation is comprised of more circuits than can be recorded on a single schedule (18 in the case of the Electrical Installation Certificate or Electrical Installation Condition Report) or the installation contains more than one distribution board or consumer unit, additional schedules will be required. For this purpose 'Continuation sheets' are available from NICEIC and ELECSA.

Continuation sheets should only be used with the current versions of certificates or report forms and should be given the same unique serial number as the certificate or report. The page number should be inserted at the foot of the sheet, together with the total number of pages comprising the certificate or report (for example, page 8 of 9). The page number(s) of the continuation sheet should be recorded in the section entitled 'Schedules and additional pages'.

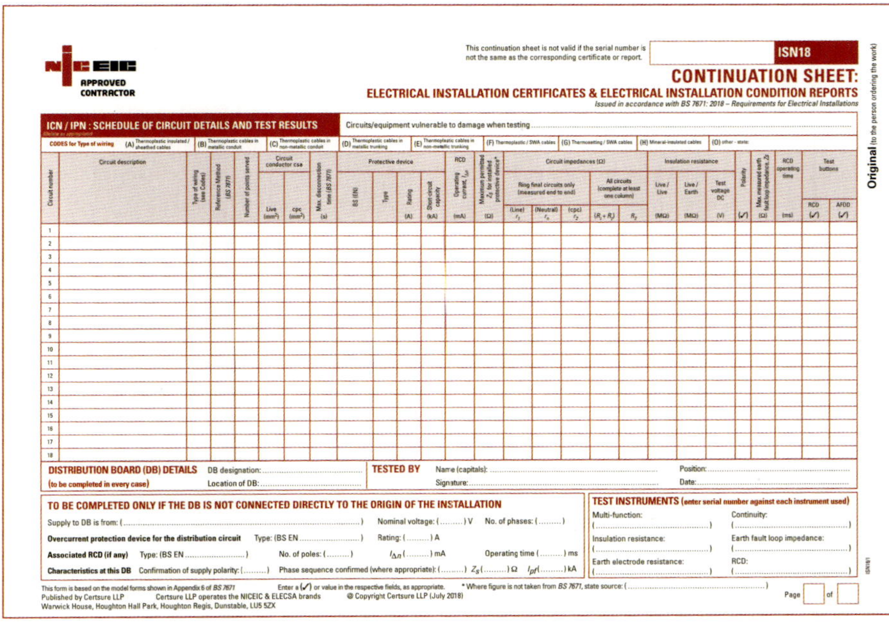

This continuation sheet is not valid if the serial number is not the same as the corresponding certificate or report.

ISN18

APPROVED CONTRACTOR

CONTINUATION SHEET:

ELECTRICAL INSTALLATION CERTIFICATES & ELECTRICAL INSTALLATION CONDITION REPORTS

Issued in accordance with BS 7671: 2018 – Requirements for Electrical Installations

Original (to the person ordering the work)

ICN / IPN : SCHEDULE OF CIRCUIT DETAILS AND TEST RESULTS

Circuits/equipment vulnerable to damage when testing ....

CODES for Type of wiring: (A) Thermoplastic insulated/sheathed cables (B) Thermoplastic cables in metallic conduit (C) Thermoplastic cables in non-metallic conduit (D) Thermoplastic cables in metallic trunking (E) Thermoplastic cables in non-metallic trunking (F) Thermoplastic / SWA cables (G) Thermosetting / SWA cables (H) Mineral-insulated cables (O) other - state:

Circuit number	Circuit description	Type of wiring (see Codes)	Reference Method (BS 7671)	Number of points served	Circuit conductor csa: Live (mm²)	Circuit conductor csa: cpc (mm²)	Max. disconnection time (BS 7671) (s)	Protective device: BS (EN)	Protective device: Type	Protective device: Rating (A)	Protective device: Short-circuit capacity (kA)	RCD: Operating current, $I_{\Delta n}$ (mA)	Maximum permitted Z_s for installed protective device* (Ω)	Circuit impedances (Ω): Ring final circuits only (measured end to end): (Line) r_1	(Neutral) r_n	(cpc) r_2	All circuits (complete at least one column): $(R_1 + R_2)$	R_2	Insulation resistance: Live / Live (MΩ)	Live / Earth (MΩ)	Test voltage DC (V)	Polarity (✓)	Max. measured earth fault loop impedance, Z_s (Ω)	RCD operating time (ms)	Test buttons: RCD (✓)	AFDD (✓)
1																										
2																										
3																										
4																										
5																										
6																										
7																										
8																										
9																										
10																										
11																										
12																										
13																										
14																										
15																										
16																										
17																										
18																										

DISTRIBUTION BOARD (DB) DETAILS (to be completed in every case) DB designation: .... Location of DB: ....

TESTED BY Name (capitals): .... Position: .... Signature: .... Date: ....

TO BE COMPLETED ONLY IF THE DB IS NOT CONNECTED DIRECTLY TO THE ORIGIN OF THE INSTALLATION

Supply to DB is from: (..........) Nominal voltage: (.........) V No. of phases: (.........)

Overcurrent protection device for the distribution circuit Type: (BS EN ....) Rating: (.........) A

Associated RCD (if any) Type: (BS EN ....) No. of poles: (.........) $I_{\Delta n}$ (.........) mA Operating time (.........) ms

Characteristics at this DB Confirmation of supply polarity: (.........) Phase sequence confirmed (where appropriate): (.........) Z_s (.........) Ω I_{pf} (.........) kA

TEST INSTRUMENTS (enter serial number against each instrument used)

Multi-function: (..........) Continuity: (..........)

Insulation resistance: (..........) Earth fault loop impedance: (..........)

Earth electrode resistance: (..........) RCD: (..........)

This form is based on the model forms shown in Appendix 6 of *BS 7671*. Enter a (✓) or value in the respective fields, as appropriate. * Where figure is not taken from *BS 7671*, state source: (..........)

Published by Certsure LLP. Certsure LLP operates the NICEIC & ELECSA brands. © Copyright Certsure LLP (July 2018)

Warwick House, Houghton Hall Park, Houghton Regis, Dunstable, LU5 5ZX

Page ☐ of ☐

ISN18/1

Fig 5.4.1 Continuation sheet for Electrical Installation Certificates and Electrical Installation Condition Reports

Where there are more observations than can be recorded in Part 6 of an Electrical Installation Condition Report (EICR) or a Domestic Electrical Installation Condition Report (DEICR), Continuation sheets are available from NICEIC and ELECSA to record any additional items relating to Part 6 'Observations and recommendations for actions to be taken'. For Electrical Installation Certificates or Electrical Installation Condition Reports the Continuation sheets should be given the same unique serial number as the Certificate or the Report, by first striking out the prefix ('DPN or IPN') as appropriate, and inserting the serial number at the top of the Continuation sheet, in the space allocated. In the case of an NICEIC or ELECSA certificate or report, the serial number should be written on each Continuation sheet.

The number of each additional page should be inserted at the foot of the page, together with the total number of pages comprising the certificate or report (e.g. page 7 of 8).

NICEIC APPROVED CONTRACTOR

This continuation sheet is not valid if the serial number is not the same as the corresponding report. IFN18

CONTINUATION SHEET FOR PART 6:

DOMESTIC ELECTRICAL INSTALLATION CONDITION REPORTS Small installations up to 100 A single phase supply & ELECTRICAL INSTALLATION CONDITION REPORTS

Issued in accordance with BS 7671: 2018 – Requirements for Electrical Installations

Original (to the person ordering the work)

DPN / IPN : OBSERVATIONS AND RECOMMENDATIONS FOR ACTIONS TO BE TAKEN
(Delete as appropriate)

CODES: *One of the following Codes, as appropriate, has been allocated to each of the observations made below to indicate to the person(s) responsible for the electrical installation the degree of urgency for remedial action*	CODE C1 'Danger Present' Risk of injury. Immediate remedial action required	CODE C2 'Potentially Dangerous' Urgent remedial action required	CODE C3 'Improvement Recommended'	CODE FI 'Further Investigation Required'

Continued from page 2 of the report

The following observations and recommendations for action are made:

Item No	Observation(s)	Code	Location Reference
(......)	(......)	(......)	(......)
(......)	(......)	(......)	(......)
(......)	(......)	(......)	(......)
(......)	(......)	(......)	(......)
(......)	(......)	(......)	(......)
(......)	(......)	(......)	(......)
(......)	(......)	(......)	(......)
(......)	(......)	(......)	(......)
(......)	(......)	(......)	(......)
(......)	(......)	(......)	(......)
(......)	(......)	(......)	(......)
(......)	(......)	(......)	(......)
(......)	(......)	(......)	(......)
(......)	(......)	(......)	(......)
(......)	(......)	(......)	(......)
(......)	(......)	(......)	(......)
(......)	(......)	(......)	(......)
(......)	(......)	(......)	(......)
(......)	(......)	(......)	(......)
(......)	(......)	(......)	(......)
(......)	(......)	(......)	(......)
(......)	(......)	(......)	(......)
(......)	(......)	(......)	(......)

Immediate action required for items: (......) **Improvement recommended** for items: (......)

Urgent remedial action required for items: (......) **Further investigation required** for items: (......)

IFN18/1

This report is based on the model forms shown in Appendix 6 of *BS 7671*

Published by Certsure LLP Certsure LLP operates the NICEIC & ELECSA brands
Warwick House, Houghton Hall Park, Houghton Regis, Dunstable, LU5 5ZX

Please see 'Guidance for recipients on the classification codes' Page [ ] of [ ]

Fig 5.4.2 Continuation sheet for Part 6 'Observations and recommendations for actions to be taken'

CHAPTER 6
THE SCHEDULE OF ITEMS INSPECTED FOR NEW INSTALLATION WORK

(i) General

Every new electrical installation, and every addition or alteration to an existing installation, should be inspected during erection and on completion before being put into service to verify, so far as is reasonably practicable, that the requirements of the *BS 7671* have been met (Regulation 641.1 refers).

(ii) Initial inspection of new work

As required by Regulation 642.2, initial inspection should be made to verify that installed electrical equipment:

- complies with the requirements of the appropriate British, European or equivalent product standards (Section 511 of *BS 7671* refers),
- is correctly selected and erected, taking into account manufacturer's instructions, and
- is not visibly damaged or defective so as to impair safety.

So that inspection, and subsequent testing, can be completed satisfactorily, all relevant schedules, drawings and design information should be made available to those carrying out such work (Regulation 641.2 refers).

Inspection should be carried out before testing, and the part of the installation being inspected should normally be disconnected from the supply (Regulation 642.1 refers). A careful and thorough inspection of installed work is vital to identify deficiencies that are unlikely to be detected by testing.

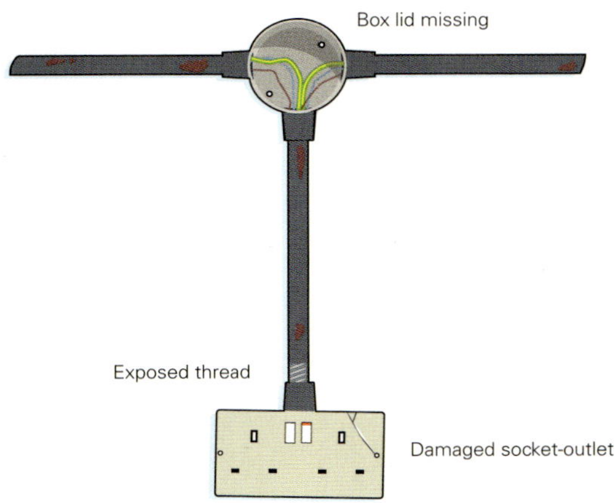

Fig 6 (i) Examples of deficiencies

As shown in Fig 6 (i), inspection should, amongst others, identify any incomplete enclosures, damaged accessories or unsuitable erection and installation methods.

It is almost always necessary for some inspection work to be carried out during the process of constructing an installation, because some of the items requiring inspection will not be accessible on completion. For example, the condition of cables concealed in the fabric of a building cannot be inspected after completion of the building work. Inspection during the construction stage will also enable work that is unsatisfactory, or work that has suffered damage, to be identified and remedied at an early stage.

It should be noted that irrespective of the type of installation, domestic, commercial or industrial, inspection should only be carried out by skilled persons competent in such work (Regulation 641.6 refers). In particular, the inspector must have sufficient experience of the particular type of installation being inspected and be competent to employ all the necessary safety precautions required.

For such reasons, an NICEIC and ELECSA certificate must only be used to certify work that is within scope of the contractor's registration.

For details on the use of NICEIC or ELECSA certificates refer to Chapter 2 and for further details on the competency of the inspector and safety precautions refer to Chapter 1.

6 The Schedule of Items Inspected

The remainder of this Chapter provides guidance on completing the 'Schedule of Items Inspected' which forms part of the NICEIC Electrical Installation Certificate (EIC). The guidance is also relevant to the modified form of this schedule which forms part of an NICEIC or ELECSA Domestic Electrical Installation Certificate (DEIC). The items listed in these schedules are not exhaustive so where appropriate, additional items may need to be included due to the type or nature of the particular installation. Where this is the case, any additional item(s) and their outcome(s) should be recorded on an additional numbered page and attached to the schedule.

All fields on the schedule should be completed. A 'tick' should be inserted to indicate that inspection, and where necessary testing, was carried out and the result was satisfactory. If an inspection item is not applicable to the particular installation then 'N/A' should be inserted. The insertion of a 'cross' to indicate that the result of an inspection was unsatisfactory, or a 'LIM', to indicate that a limitation prevented an inspection being carried out, are **not** options for an inspection schedule forming part of an EIC or DEIC.

APPROVED CONTRACTOR

This certificate is not valid if the serial number has been defaced or altered

ICN18

ELECTRICAL INSTALLATION CERTIFICATE

Issued in accordance with BS 7671: 2018 – Requirements for Electrical Installations

Original (to the person ordering the work)

PART 9 : SCHEDULE OF ITEMS INSPECTED – continues on next page

1. External condition of electrical intake equipment (visual inspection only)

1.1 Service cable: (.........) 1.2 Service head: (.........)
1.3 Earthing arrangement: (.........) 1.4 Meter tails: (.........)
1.5 Metering equipment: (.........) 1.6 Isolator (where present): (.........)

2. Parallel or switched alternative sources of supply

2.1 Presence of adequate arrangements where generator to operate as a switched alternative:
- a) Dedicated earthing arrangement independent of that of the public supply (.........)

2.2 Presence of adequate arrangements where generator to operate in parallel with public supply:
- a) Correct connection of generator in parallel (.........)
- b) Compatibility of characteristics of means of generation (.........)
- c) Means to provide automatic disconnection of generator in the event of loss of public supply or voltage or frequency deviation beyond declared values (.........)
- d) Means to prevent connection of generator in the event of loss of public supply or voltage or frequency deviation beyond declared values (.........)
- e) Means to isolate generator from public supply (.........)

2.3 Presence of alternative / additional supply warning notices at or near:
- a) The origin (.........)
- b) The meter position, if remote from origin (.........)
- c) The consumer unit / distribution board to which the alternative / additional sources are connected (.........)
- d) All points of isolation of ALL sources of supply (.........)

3. Automatic disconnection of supply

3.1 Presence and adequacy of protective earthing / bonding arrangements as follows:
- a) Distributor's earthing arrangement or installation earth electrode arrangement (.........)
- b) Earthing conductor and connections (.........)
- c) Main protective bonding conductors and connections (.........)
- d) Earthing / bonding labels at all appropriate locations (.........)

3.2 Accessibility of:
- a) Earthing conductor connections (.........)
- b) All protective bonding connections (.........)

3.3 FELV – requirements satisfied: (.........)
3.4 Reduced low voltage – requirements satisfied: (.........)

4. Additional protection

4.1 The presence and effectiveness of additional protection methods used, as follows: (.........)
- a) RCDs not exceeding 30 mA operating current, as specified (.........)
- b) Supplementary bonding (.........)

5. Basic protection *(‡ For use in controlled / supervised conditions only)*

5.1 Presence and adequacy of protective measures to provide basic protection:
- a) Insulation of live parts (.........)
- b) Barriers or enclosures (.........)
- c) Obstacles ‡ (.........)
- d) Placing out of reach ‡ (.........)

6. Basic and fault protection
- a) SELV (.........)
- b) PELV (.........)
- c) Double or reinforced insulation (.........)

When used, provide details on a separate numbered page: Page No (.........)

7. Distribution equipment

7.1 Adequacy of working space / accessibility: (.........)
7.2 Security of fixing: (.........)
7.3 Insulation of live parts not damaged during erection: (.........)
7.4 Adequacy / security of barriers: (.........)
7.5 Suitability of enclosures for IP and fire ratings: (.........)
7.6 Enclosures not damaged during installation: (.........)
7.7 Presence and effectiveness of obstacles: (.........)
7.8 Presence and operation (functional) check of main switch(es): (.........)
7.9 Components are suitable according to assembly manufacturer's instructions or literature: (.........)
7.10 Operation of circuit-breakers and RCDs to prove functionality: (.........)
7.11 RCD(s) provided for fault protection, where specified: (.........)
7.12 RCD(s) provided for protection against fire, where specified: (.........)
7.13 RCD(s) provided for additional protection, where specified: (.........)
7.14 Confirmation overvoltage protection (SPDs) provided, where specified: (.........)
7.15 Indication of SPD(s) continued functionality confirmed: (.........)
7.16 Selection of protective devices(s) and base(s); correct type and rating: (.........)
7.17 Single-pole protective devices in line conductors only: (.........)
7.18 Protection against mechanical damage where cables enter equipment: (.........)
7.19 Protection against electromagnetic effects where cables enter ferromagnetic enclosures: (.........)
7.20 Confirmation that ALL conductor connections, including connections to busbars, are correctly located in terminals and are tight and secure: (.........)
7.21 Presence of RCD six-monthly test notice, where required: (.........)
7.22 Presence of diagrams, charts or schedules at or near each distribution board, where required: (.........)
7.23 Presence of next inspection recommendation label: (.........)
7.24 Presence of non-standard (mixed) cable colour warning notice at or near the appropriate distribution board, where required: (.........)
7.25 Presence of other required labelling: (.........)

8. Circuits

8.1 Identification of conductors: (.........)
8.2 Cables correctly supported throughout, with protection against abrasion: (.........)
8.3 Examination of cables for signs of mechanical damage during installation: (.........)
8.4 Examination of installation of live parts, not damaged during erection: (.........)
8.5 Non-sheathed cables protected by enclosure in conduit, ducting or trunking: (.........)
8.6 Suitability of containment systems (including flexible conduit): (.........)
8.7 Correct temperature rating of cable insulation: (.........)
8.8 Adequacy of cables for current-carrying capacity with regard to the type and nature of installation: (.........)
8.9 Adequacy of protective devices: type and fault current rating for fault protection: (.........)
8.10 Adequacy of AFDD(s), where specified: (.........)
8.11 Presence and adequacy of circuit protective conductors: (.........)
8.12 Coordination between conductors and overload protective devices: (.........)

This certificate is based on the model forms shown in Appendix 6 of *BS 7671* Enter a (✓) or value in the respective fields, as appropriate. Where an item is not applicable insert N/A
Published by Certsure LLP Certsure LLP operates the NICEIC & ELECSA brands © Copyright Certsure LLP (July 2018)
Warwick House, Houghton Hall Park, Houghton Regis, Dunstable, LU5 5ZX

Page 4 of

ICN18/7

Fig 6 Schedule of Items Inspected

6.1 External condition of electrical intake equipment *(visual inspection only)*

Aspects of the supply intake equipment listed in this section of the schedule, which are readily accessible, should be visually inspected for signs of mechanical damage, evidence of overheating and other forms of deterioration. The metering equipment and any connection boxes provided by the distributor as a part of their installation should be securely fixed with all outer casings/enclosures undamaged.

Where a switch is provided by the distributor as a means of main isolation, it should be confirmed that the switch is functional and its condition satisfactory.

1. External condition of electrical intake equipment (visual inspection only)

1.1 Service cable: (.........) 1.2 Service head: (.........)
1.3 Earthing arrangement: (.........) 1.4 Meter tails: (.........)
1.5 Metering equipment: (.........) 1.6 Isolator (where present): (.........)

Fig 6.1.1 External condition of electrical intake equipment

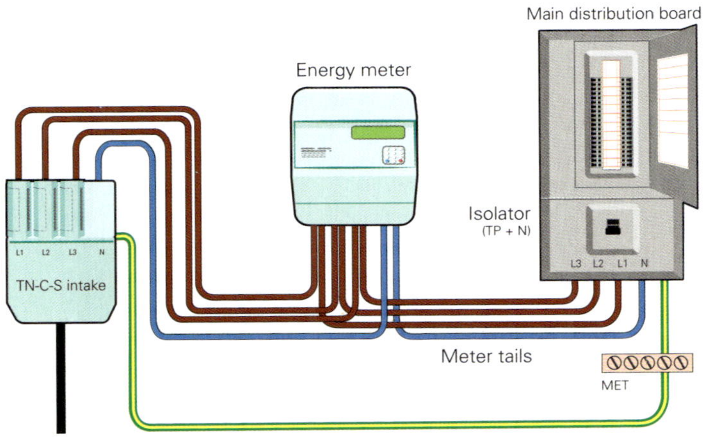

Note: Where inadequacies in the electricity distributor's or supplier's equipment are observed, the person ordering the inspection should be advised to inform the relevant authority.

Fig 6.1.2 Three-phase electrical intake

6.2 Parallel or switched alternative sources of supply

Where a supply to the installation is available from more than one source, the inspector should check that adequate arrangements in terms of protection for safety are present, including appropriate provision for isolation, manual and automatic switching, and any necessary interlocking. Specific requirements for low voltage generating sets are contained in Chapter 55 of *BS 7671*.

The items listed in this section of the schedule cover two distinct methods of operation. A generator may be installed so that it operates in the event of loss of the normal public supply (switched alternative), or it may be installed to operate in parallel with the public supply.

Where more than one supply source is available to an installation the prospective fault current (I_{pf}) for each independent source should be assessed (Regulation 551.2.2 refers). For an NICEIC or ELECSA electrical certificate, the details of each additional source should be appended to the certificate, and the higher or highest value of I_{pf} and external impedance (Z_e) should be recorded on the certificate under 'Nature of supply parameters'.

2. Parallel or switched alternative sources of supply

2.1 Presence of adequate arrangements where generator to operate as a switched alternative:

- a) Dedicated earthing arrangement independent of that of the public supply (.........)

2.2 Presence of adequate arrangements where generator to operate in parallel with public supply:

- a) Correct connection of generator in parallel (.........)
- b) Compatibility of characteristics of means of generation (.........)
- c) Means to provide automatic disconnection of generator in the event of loss of public supply or voltage or frequency deviation beyond declared values (.........)
- d) Means to prevent connection of generator in the event of loss of public supply or voltage or frequency deviation beyond declared values (.........)
- e) Means to isolate generator from public supply (.........)

Fig 6.2.1 Parallel or switched alternative sources of supply

Presence of adequate arrangements where generator to operate as a switched alternative

A generator that operates as a switched alternative to the public supply (standby generator) must be earthed independently from the public supply earthing arrangement (Regulation 551.4.3.2.1 refers), and the installation should comply with the relevant requirements of Regulation Group 551.6.

Whilst it is permitted to insert a switching device in the connection between the neutral point and the means of earthing, to ensure that only one means of earthing is applied at any one time, where such a device is used it must satisfy the requirements of Regulation 543.3.3.101.

Presence of adequate arrangements where generator to operate in parallel with the public supply

Where a generator, including a microgenerator, is intended to operate in parallel with the public supply system the additional requirements of Regulation Group 551.7 must be satisfied.

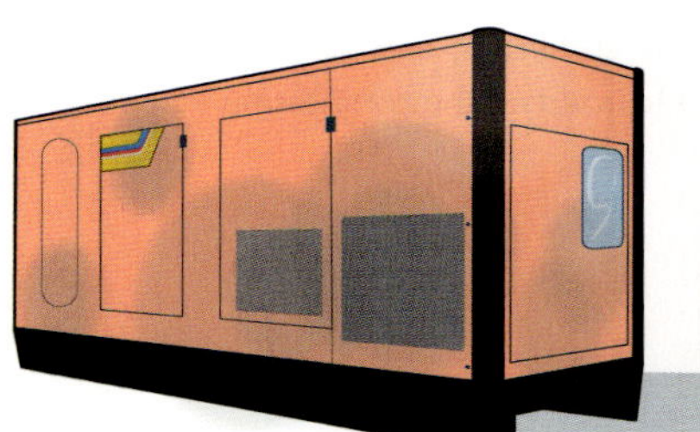

Fig 6.2.2 Standby generator

Inspection should confirm that thermal and overcurrent protection are provided in accordance with Chapter 42 and Chapter 43 of *BS 7671* and remain effective in all situations (Regulation 551.7.1 refers). Where an RCD provides additional protection in accordance with Regulation 415.1.1 for the circuit connecting the generator set to the installation, the device should disconnect all live conductors (line and neutral).

The following inspections are included on the schedule:

- **Correct connection of generator in parallel:** Inspection is required to confirm that the generating set is connected to the supply side of all overcurrent protective devices for the final circuits of the installation. If the generating set is connected to the load side of all overcurrent protective devices for the final circuits of the installation, all the requirements of Regulation 551.7.2 must be satisfied.
- **Compatibility of characteristics the of means of generation:** The characteristics of the generator should be checked to verify that its operation will not adversely affect the normal operating parameters of the installation such as, voltage, frequency, system balance, power factor and the like (Regulation 551.7.3 refers).
- **Means to provide automatic disconnection of generator in the event of loss of public supply or voltage or frequency deviation beyond declared values:** A generator operating in parallel with the public supply must disconnect automatically on loss of the supply otherwise persons who may be carrying out maintenance or repairs on the distribution network may be at risk of electric shock (Regulation 551.7.4 refers). In addition, generator settings should be checked to confirm that automatic disconnection is provided in the event of deviations to the generator's declared values of voltage or frequency.

Regulation 551.7.5 requires that a means is provided to prevent connection of a generator in the event of the circumstances described above, and that a means has been provided to isolate the generator from the public supply (Regulation 551.7.6 refers).

Particular requirements for the installation of solar photovoltaic (PV) power supply systems are contained in Section 712 of *BS 7671*.

Fig 6.2.3 A solar photovoltaic installation

Presence of alternative / additional supply warning notices at or near:

a) The origin (.........)

b) The meter position, if remote from origin (.........)

c) The consumer unit / distribution board to which the alternative / additional sources are connected (.........)

d) All points of isolation of ALL sources of supply (.........)

Fig 6.2.4 Presence of alternative / additional supply warning notices

Presence of alternative / additional supply warning notices

Where an installation includes an additional / alternative source of supply, Regulation 514.15.1 requires that a warning notice, such as that shown in Fig 6.2.5, is affixed at the following locations in the installation:

- At the installation's origin
- At the meter position, where this is remote from the origin
- At the consumer unit or distribution board to which alternative or additional sources are connected
- At all points of isolation of all sources of supply.

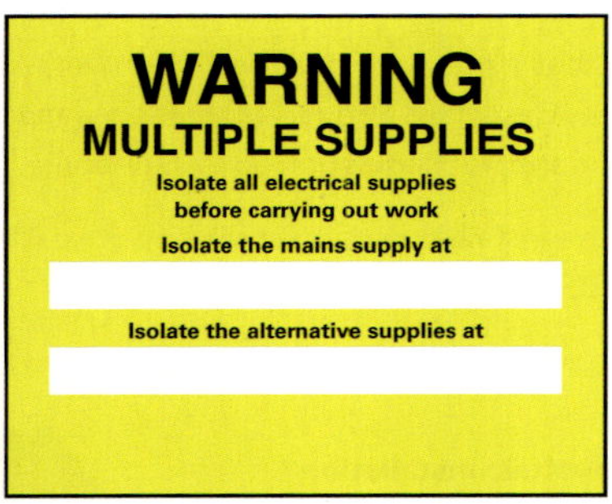

Fig 6.2.5 Warning notice to identify that the installation is connected to multiple supplies

6.3 Automatic disconnection of supply (ADS)

The protective measure ADS is used in most, if not all, electrical installations, and is a protective measure in which fault protection is provided by protective earthing, protective equipotential bonding and automatic disconnection in case of fault (Regulation 411.1 refers). Therefore, wherever ADS is employed, the inspector is required to verify the presence and adequacy of the protective earthing and bonding arrangements.

3. Automatic disconnection of supply

3.1	Presence and adequacy of protective earthing / bonding arrangements as follows:	
	a) Distributor's earthing arrangement or installation earth electrode arrangement	(.........)
	b) Earthing conductor and connections	(.........)
	c) Main protective bonding conductors and connections	(.........)
	d) Earthing / bonding labels at all appropriate locations	(.........)
3.2	Accessibility of:	
	a) Earthing conductor connections	(.........)
	b) All protective bonding connections	(.........)
3.3	FELV – requirements satisfied:	(.........)
3.4	Reduced low voltage – requirements satisfied:	(.........)

Fig 6.3.1 Automatic disconnection of supply

Distributor's earthing arrangement or installation earth electrode arrangement

The earthing arrangement should be checked to confirm that it is suitable for the particular supply system (Regulation 542.1.201 refers). Where an installation earth electrode is installed as the means of earthing for the installation, compliance with the relevant requirements of Regulation Group 542.2 should be confirmed. Recognised types of earth electrode are listed in Regulation 542.2.2. Where an electrode is installed a measurement of the electrode's resistance to Earth is required see Chapter 8 'Testing'.

Earthing conductor and connections

The earthing conductor is the protective conductor which connects the Main Earthing Terminal (MET) of the installation to the means of earthing. The earthing conductor is a vitally important part of the earth fault loop path, so it must be confirmed that the earthing conductor is present and its connection to the means of earthing is electrically and mechanically sound and suitably protected against corrosion (Regulation 542.3.2 refers).

The cross-sectional area of the earthing conductor should comply with Section 543, and except where PME conditions apply it should satisfy the protective bonding requirements given in Regulation 544.1.1.

The minimum cross-sectional area of the earthing conductor may have been determined either by calculation, or by selection in accordance with Table 54.7 of *BS 7671*. However, where the earthing conductor is buried in the ground, as is generally the case where the system earthing is TT, its cross-sectional area must not be less than that given in Table 54.1 of *BS 7671*.

If the cross-sectional area of the earthing conductor is less than that indicated in Table 54.7 (or Table 54.1, if applicable), the designer should be consulted to confirm the adequacy of the earthing conductor arrangement.

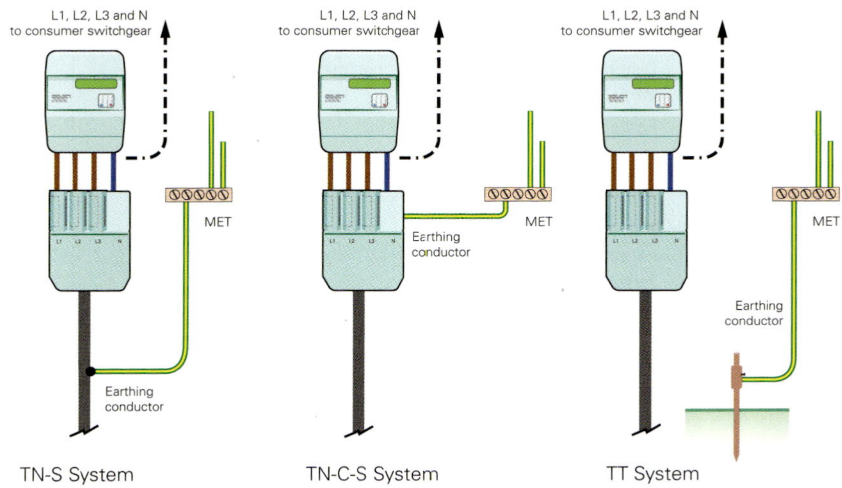

Fig 6.3.2 Earthing arrangements

Main protective bonding conductors and connections

A key component of the protective measure ADS is that extraneous-conductive-parts should be connected to the main earth terminal (MET) of the particular installation using protective bonding conductors complying with Chapter 54 of *BS 7671* (Regulation 411.3.1.2 refers).

Main protective bonding helps to ensure that the magnitude of voltages occurring between simultaneously accessible exposed-conductive-parts and extraneous-conductive-parts under earth fault conditions is insufficient to cause danger during the time it takes for the relevant protective device to achieve disconnection of the faulty circuit.

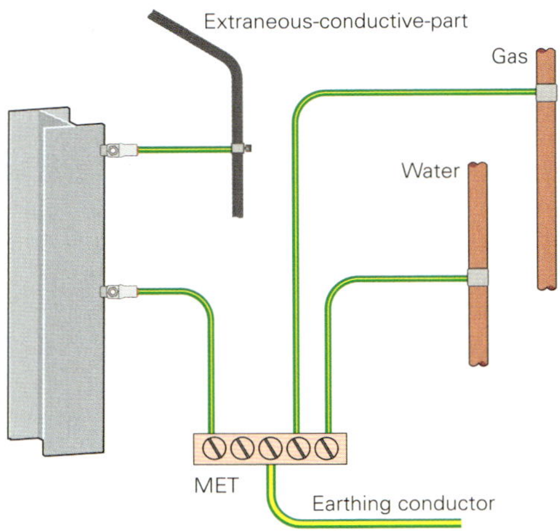

(Earthing labels omitted for clarity)

Fig 6.3.3 Main protective bonding to extraneous-conductive-parts

While Regulation 411.3.1.2 provides a list of items that require protective bonding, including, among others, metallic pipework and structural metalwork, it should be appreciated that this list is neither exhaustive nor can it be assumed that all the items identified will always fall within the definition of an extraneous-conductive-part. As stated in Regulation 411.3.1.2, a metallic pipe comprised of an insulating section on entry into the building is not deemed an extraneous-conductive-part and therefore, does not require bonding (for details on extraneous-conductive-parts refer to Appendix E of this Guide).

Except where Protective Multiple Earthing (PME) conditions apply, the cross-sectional area of main protective bonding conductors must be not less than half the cross-sectional area required by *BS 7671* for the earthing conductor of the installation, with a minimum of 6 mm^2. The cross-sectional area need not exceed 25 mm^2 if the bonding conductor is of copper. If a copper bonding conductor is not used, its cross-sectional area must be such that its conductance is not less than that of the required size of copper conductor (Regulation 544.1.1 refers).

For an installation connected to a PME supply (a TN-C-S system), with the exception of highways power supplies and street furniture, the cross-sectional area of the main protective bonding conductors must be not less than that given in Table 54.8 of *BS 7671*. The values from Table 54.8 are shown in Table 6.3.1. It should be noted, however, that in some instances the electricity distributor may require a larger cross-sectional area than the minimum values stated.

Table 6.3.1 Minimum cross-sectional area (csa) of the main protective bonding conductor in relation to the PEN conductor of the supply

Copper equivalent csa of the PEN conductor (in mm^2)	Minimum copper equivalent csa of the main protective bonding conductor (in mm^2)
35 or less	10
over 35 up to 50	16
over 50 up to 95	25
over 95 up to 150	35
over 150	50

Note: Where the bonding conductor is not of copper, the conductor material that is used must afford the equivalent conductance of the minimum cross-sectional area required for a copper bonding conductor.

Earthing/bonding labels at all appropriate locations

As required by Regulation 514.13.1, a permanent label to *BS 951* with the words, 'Safety Electrical Connection - Do Not Remove' must be permanently fixed in a visible position at or near:

- The connection of an earthing conductor to an installation earth electrode.
- The connection of a protective bonding conductor to an extraneous-conductive-part.
- The main earthing terminal, where it is separate from the main switchgear.

Accessibility of earthing/bonding connections

Earthing and protective bonding connections should be accessible for the purposes of inspection, testing and maintenance (Regulation 543.3.2 refers).

The connection of a main protective bonding conductor to an extraneous-conductive-part such as metallic gas, water or other service pipe should be carried out as near as practicable to the point at which the part enters the premises. The connection being made on the hard installation pipework, before any branch on the consumer's side of any meter, isolation point or union.

Where practicable, the protective earth bond (PEB) connection should be made within 600 mm of the meter outlet union. Where the meter is external, the bonding connection should be made at the point(s) of entry of the service into the building (Regulation 544.1.2 refers).

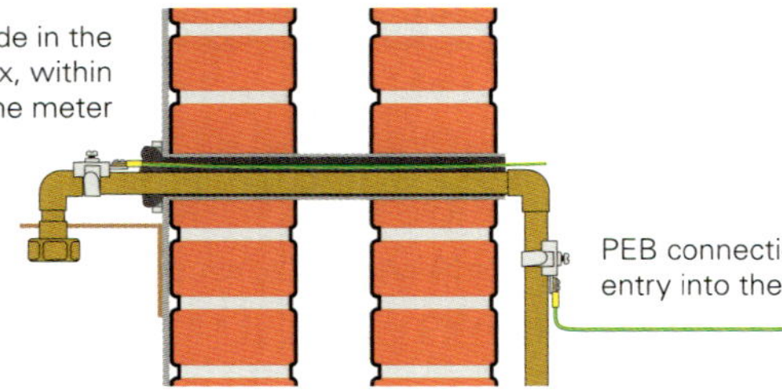

Fig 6.3.4 Protective earth bond (PEB) connection made either internally or externally

Functional extra-low voltage (FELV) – requirements satisfied

A functional extra-low voltage system is a system where not all of the protective measures required for SELV and PELV have been applied. A FELV system may be used for both basic and fault protection provided that:

- basic protection is provided by basic insulation of live parts and/or barriers or enclosures, and
- fault protection is achieved by ensuring the exposed-conductive-parts of the equipment of the FELV circuit are connected to the protective conductor of the primary circuit of the source. The primary circuit having automatic disconnection of supply.

Where FELV is used, the inspector should confirm that the protective provisions for basic and fault protection and the other requirements detailed in Section 411.7 have been provided.

Reduced low voltage (RLV) – requirements satisfied

Where an RLV system is installed the inspector should confirm the requirements for fault and basic protection are satisfied in accordance with Section 411.8. The source of supply to a RLV circuit should comply with Regulation 411.8.4.1, and any connecting method, such as a plug, socket or other connecting device connected to an RLV circuit, must not be dimensionally compatible with those of any other system present in the same premises (Regulation 411.8.5 refers).

Conductors of cables used in RLV systems, as for any other conductors, should be identified in accordance with Table 51 of *BS 7671* (refer to 'Identification of conductors' item 8 of this Chapter).

6.4 Additional protection

4. Additional protection

4.1	The presence and effectiveness of additional protection methods used, as follows:	(.........)
	a) RCDs not exceeding 30 mA operating current, as specified	(.........)
	b) Supplementary bonding	(.........)

Fig 6.4.1 Additional protection

RCDs not exceeding 30 mA operating current, as specified

Where additional protection is provided by an RCD for AC systems, each RCD provided for additional protection should be inspected to confirm that it has a rated residual operating current ($I_{\Delta n}$) not exceeding 30 mA (Regulation 415.1.1 refers).

Subsequently each device should be tested at a current exceeding its residual operating current to confirm that it operates within 40 ms. Requirements for the provision of additional protection are covered later in this Chapter in item 8.

Inspection should also confirm that the number of points (or circuits) controlled by a single RCD are such that the operation of that RCD would not result in unreasonable hazards or inconvenience (Regulation 314.1 and Regulation 531.3.2 refer).

Note: An RCD must not be used as the sole means of protection against electric shock (see Regulation 415.1.2), one or more of the protective measures covered in this Chapter must also be provided.

Supplementary protective equipotential bonding (Supplementary bonding)

The requirement for supplementary bonding is generally applicable to installations and locations where there is an increased risk of electric shock. Supplementary bonding may be applied for the whole installation, a particular location or an item of apparatus. The provision of supplementary bonding is required as additional protection for several of the locations and installations covered by Part 7 of *BS 7671*. For a location containing a shower or bath supplementary bonding may be omitted if the conditions detailed in Regulation 701.415.2 are fulfilled.

Except where Regulation 419.2 applies, where automatic disconnection cannot be achieved by a protective device, supplementary protective equipotential bonding should be provided in accordance with Regulation Group 415.2, and the voltage between simultaneously accessible conductive parts shall not exceed 50 V AC or 120 V DC (Regulation 419.3 refers).

Where a supplementary bonding conductor is installed, compliance with the relevant requirements of Regulation Group 544.2 should be confirmed.

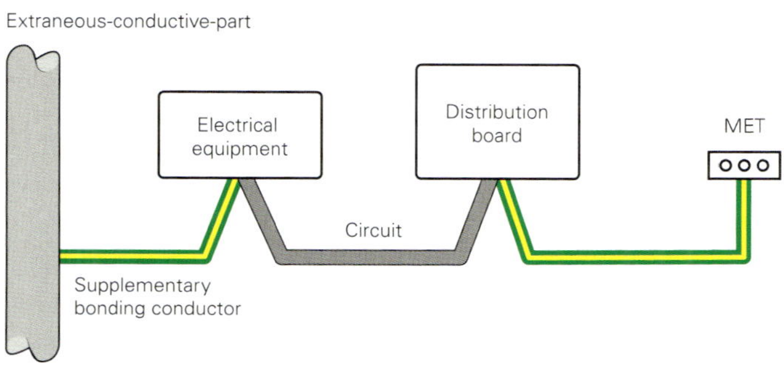

Fig 6.4.2 Supplementary protective equipotential bonding

Note: A supplementary bonding conductor that is **not** provided with mechanical protection should have a minimum cross-sectional area of 4 mm^2.

6.5 Basic Protection

5. Basic protection *(‡ For use in controlled / supervised conditions only)*

5.1 Presence and adequacy of protective measures to provide basic protection:

a)	Insulation of live parts	(.........)
b)	Barriers or enclosures	(.........)
c)	Obstacles ‡	(.........)
d)	Placing out of reach ‡	(.........)

Fig 6.5.1 Basic protection

Basic protection is intended to prevent contact with live parts and the items listed in this require confirmation of the presence and adequacy of protective measures used to provide basic protection. Provisions for basic protection are given in Section 416 of *BS 7671*.

Insulation of live parts

Inspection of this protective provision includes checking that live parts are completely covered with insulation. When inspecting equipment, the insulation must comply with the relevant standard for such electrical equipment. Paint, varnish, lacquer or similar products are g not considered to provide adequate insulation for basic protection in normal service.

Barriers or enclosures

Barriers or enclosures are intended to prevent contact with live parts. A barrier p protection from contact with live parts from any usual direction of access whereas an en provides protection from contact with live parts from any direction.

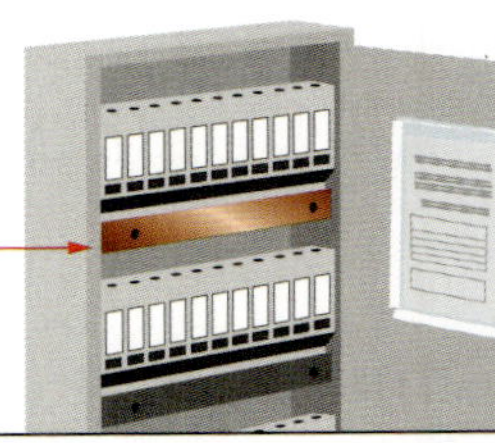

Fig 6.5.2 (a) Protection by barriers

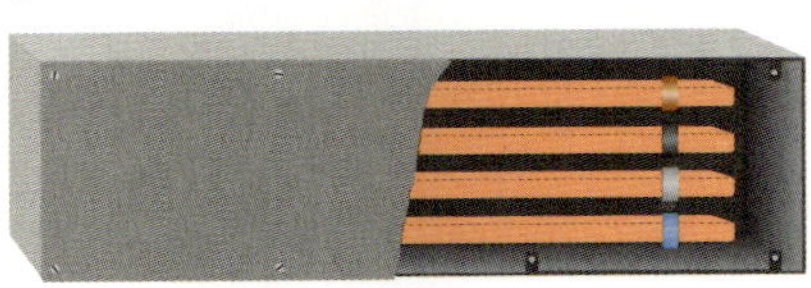

Fig 6.5.2 (b) Protection by enclosures

Inspection should verify that barriers and enclosures provide the minimum degree of protection. For example, all unused entries in enclosures should be closed, and no blanks should be missing from spare ways in consumer units or distribution boards (Regulation 416.2.4 refers). An understanding of *BS EN 60529: 1992 +A2: 2013 Specification for degrees of protection provided by enclosures* (IP code), is required for the inspection of barriers and enclosures. A table of IP codes is provided in Appendix F of this Guide.

Apart from certain exceptions, barriers or enclosures containing live parts must provide a degree of protection of at least IP2X or IPXXB (Regulation 416.2.1 refers). The horizontal top surface of a barrier or enclosure which is readily accessible must provide a degree of protection of at least IP4X or IPXXD (Regulation 416.2.2 refers).

Note: The letter 'X' indicates that no particular degree of protection has been specified. It does not necessarily mean that no such protection is provided.

Obstacles

An obstacle is defined by *BS 7671* as; 'a part preventing unintentional contact with live parts but not preventing deliberate contact'. An obstacle, which may be as simple as a set of railings, may be removed without the use of a tool or key, but it should be secured so as to prevent unintentional removal.

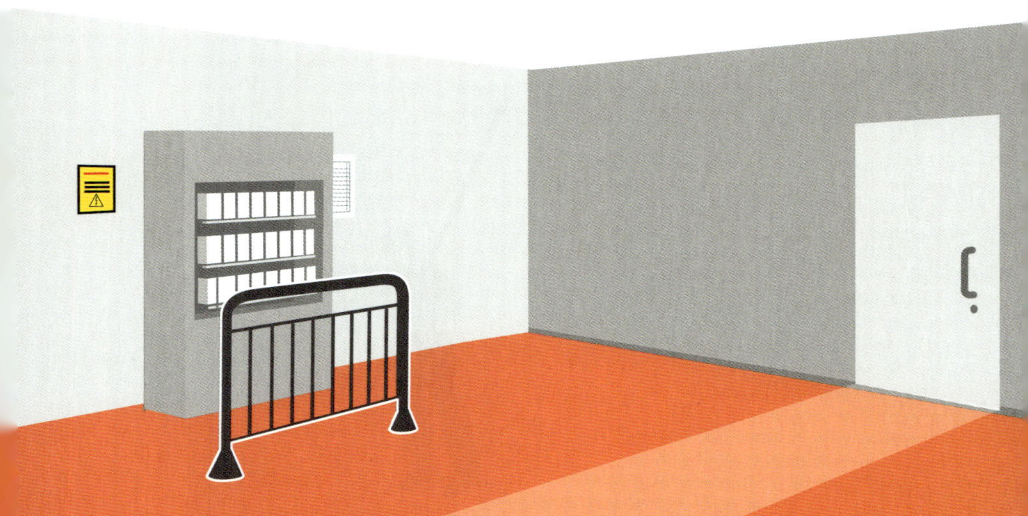

Fig 6.5.3 An obstacle placed to prevent inadvertent contact

As indicated on the 'Schedule of Items Inspected' this protective measure should only be used in controlled or supervised conditions, such as in locations where access is restricted to persons under the supervision of a skilled person. The presence of warning notices associated with such restriction must be verified.

Placing out of reach

This protective measure is intended to prevent unintentional contact with a bare live part by placing it out of reach. The measure may be used only in locations which are under the control of skilled persons. The presence of such restrictions must be verified, along with the minimum clearance distances required to satisfy the definition of 'arm's reach' given in Regulation Group 417.3 and Fig 417 of *BS 7671*. It should be noted that 'arm's reach' refers to an overhead clearance distance of 2.5 metres and applies to direct contact without assistance (e.g. tools or ladder). Therefore, where long or bulky objects, such as aluminium ladders or scaffold poles, are likely to be handled in the location and there is a risk of contact with live parts, the designer should have allowed for suitably increased clearance distances, and the continued maintenance of such increased distances must be verified by inspection.

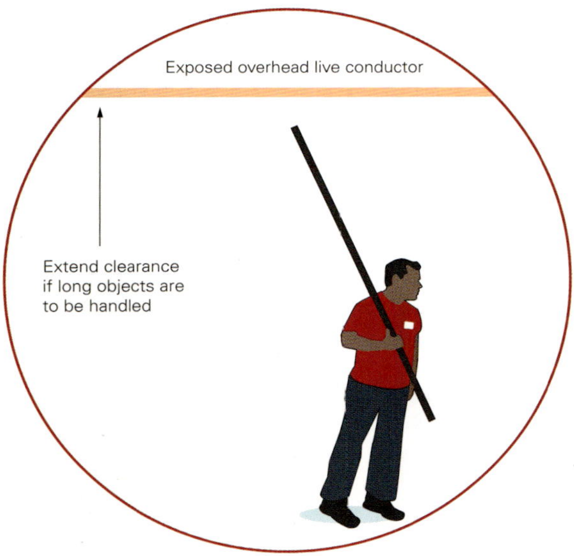

Fig 6.5.4 Placing out of reach

6.6 Basic and fault protection

The protective measure 'extra-low voltage' may consist of either separated extra-low voltage (SELV) or protective extra-low voltage (PELV), and where these systems are employed the relevant requirements of Section 414 must be satisfied.

6. Basic and fault protection	
a) SELV	(.........)
b) PELV	(.........)
c) Double or reinforced insulation	(.........)
When used, provide details on a separate numbered page:	Page No (.........)

Fig 6.6.1 The presence and effectiveness of methods which give both basic and fault protection

Basic and fault protection are deemed to be provided for SELV and PELV systems where:

- the nominal voltage source cannot exceed the upper limit of Voltage Band I, and
- the supply is from a source listed in Regulation 414.3, and
- the conditions of Regulation 414.4 are fulfilled.

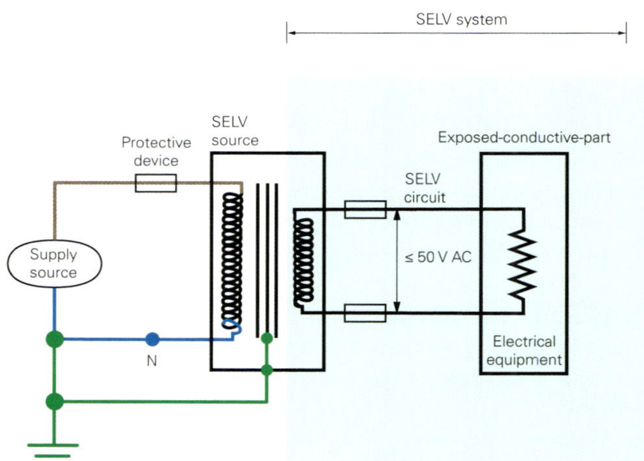

Fig 6.6.2 (a) SELV

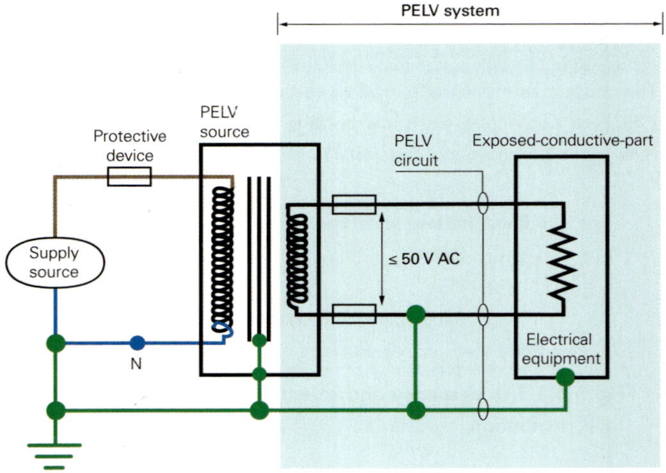

Fig 6.6.2 (b) PELV

Where, as is commonly the case, the supply source is a safety isolating transformer, it should be a type complying with *BS EN 61558-2-6* or *BS EN 61558-2-8.*

It should be noted that SELV systems are electrically separated from Earth and from other systems, so the exposed-conductive-parts of SELV circuits MUST not be connected to earth, or to any protective conductor or exposed-conductive-part part of any other circuit (Regulation 414.4.4 refers). However, a PELV circuit and/or exposed-conductive-parts of equipment supplied by a PELV circuit may be earthed (Regulation 414.4.1 refers) and as such will not be suitable in situations where a SELV system has been specified, such as for certain zones of a swimming pool.

In addition to confirming that every socket-outlet and luminaire supporting coupler of a SELV or PELV system is not dimensionally compatible with those used for other systems, the inspector must also confirm that the plugs and socket-outlets of SELV circuits have no protective conductor contact (Regulation 414.4.3 refers).

Double or reinforced insulation

This protective measure is intended to prevent a dangerous voltage appearing on accessible parts of electrical equipment through a fault in basic insulation. Basic protection is provided by basic insulation of live parts, and fault protection is provided by supplementary insulation, or both basic and fault protection are provided by reinforced insulation between live parts and accessible parts.

Where double or reinforced insulation is to be used as the sole protective measure (for example where the whole installation or circuit is intended to consist entirely of equipment with double insulation or reinforced insulation), it must be verified that the installation or circuit concerned will be under effective supervision in normal use (Regulation 412.1.2 refers).

This is to prevent changes being made to the installation or circuit that could impair the effectiveness of the protective measure. The protective measures of double insulation and reinforced insulation must not be applied to any circuit that includes a socket-outlet, luminaire supporting coupler (LSC), a device for connecting a luminaire (DCL) or cable coupler, where the user might change items of equipment without authorisation.

Where a Class I consumer unit is installed in an installation forming part of a TT system and the outgoing circuits of the consumer unit are provided with RCD protection, as is generally the case for new installations in domestic premises, cables supplying the consumer unit (meter tails) should have protection equivalent to reinforced or double insulation. (Refer later in this Chapter to 'RCD(s) provided for fault protection, where specified'.)

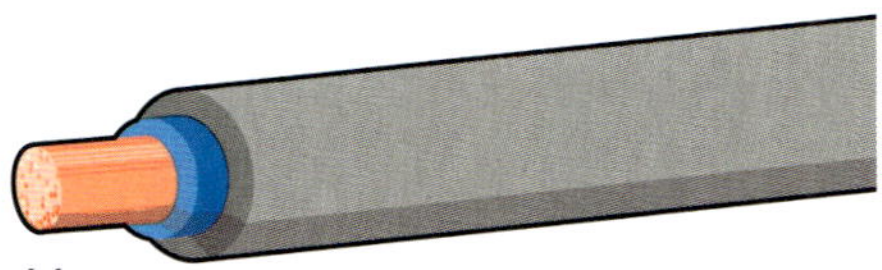

Fig 6.6.3 Double insulated cable

Electrical separation for one item of equipment

Electrical separation is a protective measure that may be applied to circuits operating at a voltage of up to 500 V, in which:

Basic protection: is provided by basic insulation of live parts or by barriers or enclosures in accordance with Section 416 of *BS 7671*. Inspection of these provisions should be carried out as discussed earlier in this Chapter.

Fault protection: is provided by simple separation of the separated circuit from other circuits and from Earth.

In order to establish that fault protection is provided the inspector must be satisfied that the requirements of Regulation Group 413.3 have been met. For example, it should be verified that:

- The separated circuit is supplied from a source of supply with at least simple separation. This could be for example a double-wound transformer the secondary winding of which is not earthed.
- That no live part is connected at any point to another circuit or to Earth, or to a protective conductor either deliberately or unintentionally, as could occur due to a damaged cable.
- Flexible cables that are liable to mechanical damage are visible throughout their length.

Also, wiring comprised of either a separate wiring system (preferably) or one of the alternatives given in Regulation 413.3.5 (such as suitable cables in insulating conduit) may be used in circumstances where the rated voltage is not less than the highest nominal voltage and overcurrent protection is provided for each circuit.

Where electrical separation is used to supply an individual item of equipment, the inspector should verify that no exposed-conductive-part of the separated circuit, such as any metal casing of an item of equipment fed by the circuit, is connected to the protective conductor or exposed-conductive-parts of another circuit, or to Earth. An example of electrical separation for one item of equipment is a shaver supply unit.

Other methods of protection

The following methods of protection are permitted only where effective supervision is provided, and their use is prohibited in certain installations and locations of increased shock risk as covered in Part 7 of *BS 7671*.

These methods are not included on the 'Schedule of Items Inspected' that forms part of NICEIC and ELECSA certificates. Therefore, if the following protective methods are employed, the details of the method and the inspections undertaken should be provided on a separate page and attached to the 'Schedule of Items Inspected'.

Non-conducting location/Earth-free local equipotential bonding

Owing to their specialised nature, these protective measures are not included within the scope of this book. Reference should be made to Section 418 of *BS 7671* for the main requirements relating to protection by non-conducting location and Earth-free equipotential bonding.

Electrical separation for the supply to more than one item of current-using equipment

Where this protective measure is used, compliance with the requirements of Regulation Group 418.3 should be confirmed. Including providing confirmation that a warning notice is fixed at every point of access to the location where the protective measure is used.

6.7 Distribution equipment

The items listed in this section refer to the installed distribution equipment such as distribution boards, consumer units, switchgear and the like. The following inspections should be carried out as appropriate to the installation.

7. Distribution equipment		
7.1	Adequacy of working space / accessibility:	(.........)
7.2	Security of fixing:	(.........)
7.3	Insulation of live parts not damaged during erection:	(.........)
7.4	Adequacy / security of barriers:	(.........)
7.5	Suitability of enclosures for IP and fire ratings:	(.........)
7.6	Enclosures not damaged during installation:	(.........)
7.7	Presence and effectiveness of obstacles:	(.........)
7.8	Presence and operation (functional) check of main switch(es):	(.........)
7.9	Components are suitable according to assembly manufacturer's instructions or literature:	(.........)

Fig 6.7.1 Distribution equipment

Adequacy of working space/accessibility

Adequate and safe means of access and working space should be provided for every piece of electrical equipment and switchgear to facilitate operation, inspection, testing, fault detection, maintenance and repair (Regulation 132.12 refers).

Where appropriate, such as in certain larger installations, the requirements of Section 729 of *BS 7671* should be confirmed. These requirements apply to operating and maintenance gangways in locations where access is restricted to skilled or instructed persons only, such as plant and intake rooms and include:

- minimum acceptable clearances that are applicable where basic protection is provided by barriers or enclosures or by obstacles
- acceptable arrangements for access to gangways
- minimum passing widths between electrical equipment and walls.

Additional requirements are included for the purposes of easy evacuation in areas of closed restricted access. As shown in Fig 6.7.2, where doors can be fixed in the open position and circuit-breakers can be withdrawn from their enclosure, a minimum passing distance of 500 mm should be maintained (Annex 729.2 of Section 729 refers).

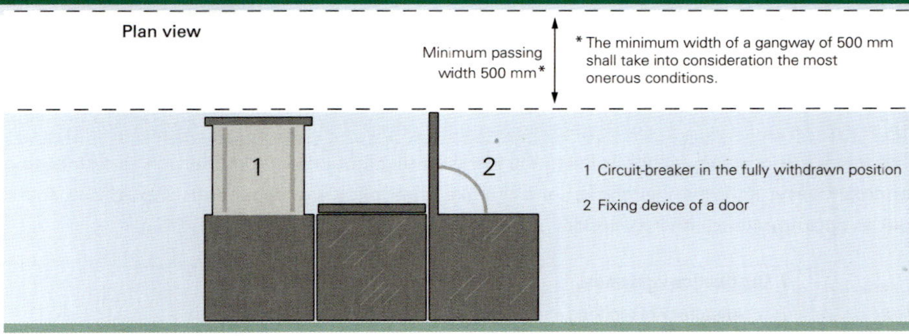

Fig 6.7.2 Adequacy of working space / accessibility

Security of fixing

Inspection should confirm that all installed equipment is securely fixed using appropriate methods and taking into account the equipment manufacturers instructions. (Regulation 134.1.1 refers).

Insulation of live parts not damaged during erection

The risk of mechanical damage and the damaging effects of external influences such as dust and water can be greater during the construction process than when the installation is in service. Inspection should, therefore, confirm that insulation intended to protect against contact with live parts has not been impaired during the construction process (Regulation 416.1 refers).

Adequacy / security of barriers

It should be verified that barriers intended to prevent contact with live parts are present and undamaged. For example, consumer units and distribution boards should be checked to confirm that blanks have been inserted in all spare ways and all unused entries have been suitably closed (Regulation 416.2 refers).

Suitability of enclosures for IP and fire ratings

Enclosures must provide the necessary minimum degree of protection against the ingress of solid objects and liquids (see Appendix F of this Guide - IP Codes). The inspector should seek to confirm that the enclosures of installed distribution equipment and switchgear comply with the design and the appropriate British or harmonised standard and are suitable for their particular location.

Where a consumer unit is installed in a domestic premises, it should be confirmed that the unit and assembly complies with *BS EN 61439-3*[1], as amended, and is made of non-combustible material, such as steel, or enclosed within a non-combustible cabinet or enclosure (Regulation 421.1.201 refers).

[1]*BS EN 61439-3:2012 Low-voltage switchgear and controlgear assemblies. Part 3: Distribution boards intended to be operated by ordinary persons (DBO)*

Enclosures not damaged during Installation

The risk of mechanical damage to enclosures is greater during the construction process than when the installation is in service. Consequently, inspection should confirm that enclosures have not been cracked, broken or otherwise impaired.

Presence and effectiveness of obstacles

Where used, the inspector should verify that the obstacle has been properly secured to prevent unintentional contact with live parts, and is located in an area that is controlled or supervised by skilled persons (see 'Obstacles' Section 5 of this Chapter).

The following items refer to the devices installed within distribution equipment such as circuit-breakers, RCDs and surge protective devices (SPDs).

Presence and operation (functional check) of main switch(es)

A main linked switch or linked circuit-breaker should be provided at or near the origin of every installation, and where the device is intended for operation by ordinary persons (such as in the case of a consumer unit installed in a domestic premises) it should interrupt all live conductors of the single-phase supply (Regulation 462.1.201 refers). Where an installation is supplied from more than one source, a main switch or main circuit-breaker is required for each source.

These devices should be checked against the design and functional testing (manual test) should be carried out to verify their operation (Regulation 643.10 refers). Information on the selection of devices for isolation and switching is provided in Table 537.4 of *BS 7671*.

Note: Isolation or switching of the neutral is not required for TN-C-S or TN-S systems if equipotential bonding is installed and either of the conditions listed in Regulation 461.2 are satisfied. However, as stated in that same regulation this does not apply to a PEN conductor.

Components are suitable according to assembly manufacturer's instructions or literature

As required by Regulation 536.4.203, mechanical and electrical devices and components, such as circuit-breakers, control devices and busbars and the like, should afford compliance with the relevant part of *BS EN 61439* series. The devices and components used in distribution boards and consumer units should be in accordance with the assembly manufacturer's instructions or literature.

7.10	Operation of circuit-breakers and RCDs to prove functionality:	(.........)
7.11	RCD(s) provided for fault protection, where specified:	(.........)
7.12	RCD(s) provided for protection against fire, where specified:	(.........)
7.13	RCD(s) provided for additional protection, where specified:	(.........)
7.14	Confirmation overvoltage protection (SPDs) provided, where specified:	(.........)
7.15	Indication of SPD(s) continued functionality confirmed:	(.........)
7.16	Selection of protective devices(s) and base(s); correct type and rating:	(.........)
7.17	Single-pole protective devices in line conductors only:	(.........)
7.18	Protection against mechanical damage where cables enter equipment:	(.........)
7.19	Protection against electromagnetic effects where cables enter ferromagnetic enclosures:	(.........)
7.20	Confirmation that ALL conductor connections, including connections to busbars, are correctly located in terminals and are tight and secure:	(.........)

Fig 6.7.3 Distribution equipment

Operation of circuit-breakers and RCDs to prove functionality

The satisfactory operation (manual) of all circuit-breakers and RCDs should be confirmed (Regulation 643.10 refers). Operation of the integral test button should cause the RCD to trip.

Protection by an RCD

Where an RCD is installed it should be selected in accordance with the relevant requirements of Regulation Group 531.3.4 and the risk of unwanted tripping should be taken into account (Regulation 531.3.2 refers). RCD protection may be installed for different purposes, such as to provide:

- fault protection
- protection against fire
- additional protection

Therefore, in addition, to any general requirements the particular requirements relating to the type of protection provided by the RCD should be confirmed.

RCD(s) provided for fault protection, where specified

Where an RCD is used to provide fault protection, compliance with the relevant requirements of Regulation Group 531.3.5 should be confirmed. This should include verifying that the type of RCD installed is compatible with the earthing arrangement and will provide the appropriate disconnection under fault conditions in accordance with the requirements of Regulation 411.3.

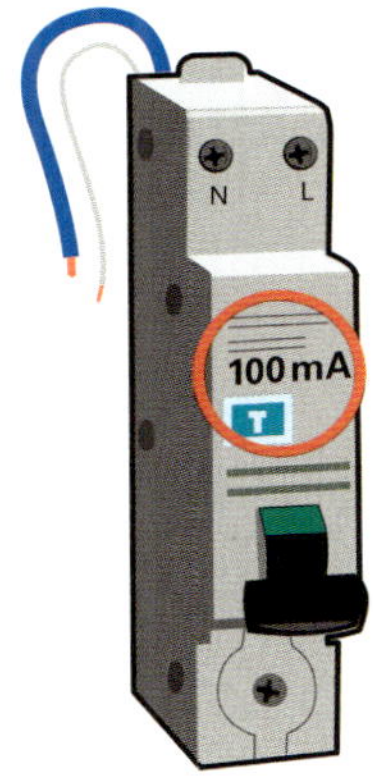

Fig 6.7.4 An RCBO rated to provide fault protection

Generally, RCDs are used to provide fault protection for the circuits of an installation connected to a TT system. Where this is the case, it should be verified that the maximum residual operating current of each RCD installed correlates with the corresponding maximum permitted value of earth resistance R_A given in Table 53.1. For example, where fault protection is provided by an RCD having a rated residual operating current of 100 mA, it should be confirmed that the maximum earth electrode resistance to Earth (R_A) does not exceed 500 Ω[1]. For guidance on measuring earth electrode resistance refer to Chapter 8.

Where a metal-clad distribution board or consumer unit is installed in an installation forming part of a TT system, an earth fault occurring between the incoming line conductor of the meter tails and the metallic enclosure, as shown in Fig 6.7.5, is unlikely to disconnect the fuse protecting the tails, due to the high impedance of the earth fault loop path. As a result, persons may be at risk of electric shock. Installing an RCD in the circuit (tails) supplying the distribution board or consumer unit to provide additional protection would not only lead to a selectivity issue (Regulation 531.3.2 'Unwanted tripping' refers) it would also still leave part of the supply (tails), between the meter and the intake of the RCD unprotected. Therefore, in such circumstances the protective measure of double or reinforced insulation should be provided on the supply side (Regulation 531.3.5.3.2.201 refers).

Where this is the case, inspection should confirm that the cables supplying the metallic consumer unit, distribution board or similar enclosure are installed so that the risk of damage is minimised. For the example shown in Fig 6.7.5, this is generally achieved where the double-insulated cables are kept as short as possible and provided with adequate protection against abrasion on entry into the consumer unit. (See 'Protection against mechanical damage where cables enter equipment'.)

[1] A value greater than 200 Ω may not be stable (Note 2 under Table 41.5 refers).

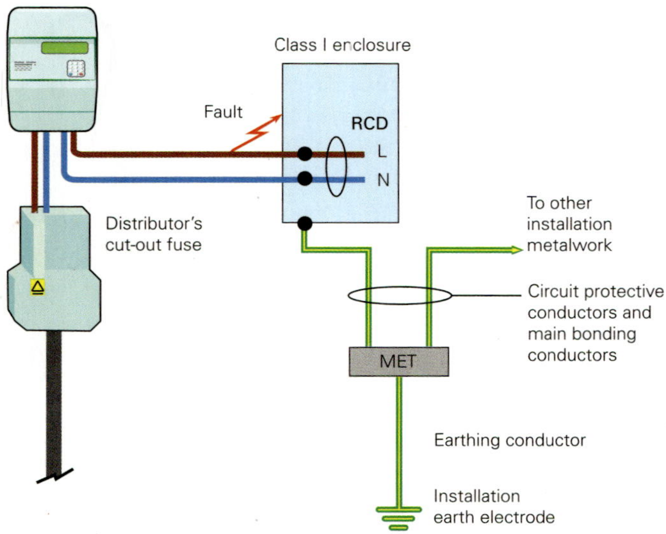

Fig 6.7.5 Risk of fault to a Class I consumer unit in an installation connected to a TT system

Note: Insulated and non-metallic sheathed cables are deemed to meet the requirements for double or reinforced insulation.

RCD(s) provided for protection against fire, where specified

An RCD used to provide protection against fire should have a residual operating current rating not exceeding 300 mA, and be installed at the origin of the circuit(s) to be protected (Regulation 532.2 refers).

Additionally, where a wiring system is installed in a location with an increased risk of fire due to the nature of processed or stored materials (Section 422.3 of *BS 7671*), it should be confirmed that the type and rating of the installed RCD satisfies the requirements of Regulation 422.3.9.

RCD(s) provided for additional protection, where specified

Where an RCD is used to provide additional protection it should have a rated residual operating current not exceeding 30 mA (Regulation 415.1.1 refers), and as required by Regulation 531.3.6, should comply with:

- *BS EN 61008* series, or
- *BS EN 61009* series, or
- *BS EN 62423.*

RCDs intended to be operated by ordinary persons should comply with Regulation 531.3.4.1 and be designed and installed so that adjustment or modification of the rated residual operating current of the RCD can only be achieved by a deliberate act (Regulation 531.3.4.201 refers). RCDs that are intended to be operated by instructed or skilled persons should comply with Regulation 531.3.4.2.

Note: An RCD with a rated residual current not exceeding 30 mA may provide both fault protection and additional protection (Regulation 531.3.6 refers). However, an RCD having a residual current rating exceeding 30 mA is not suitable for additional protection.

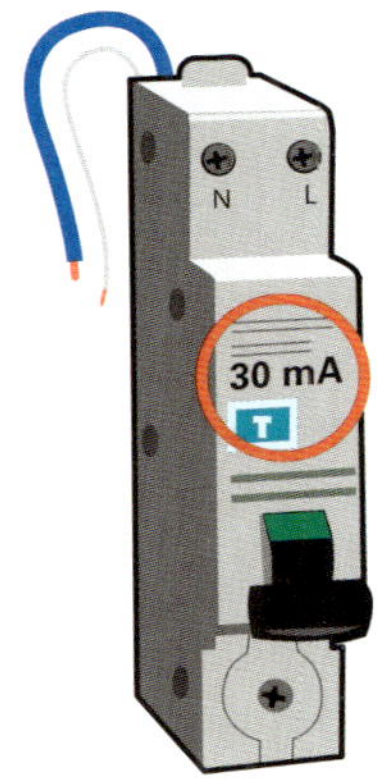

Fig 6.7.6 An RCBO rated to provide additional protection

Unwanted tripping

RCDs should be selected and erected so that the risk of unwanted tripping is minimised. In view of this, Regulation 531.3.2 requires the following factors to be taken into account:

- The appropriate division of circuits, in accordance with Section 314 of *BS 7671*, so that any earth leakage currents occurring during normal operation do not cause unwanted tripping of RCDs.
- Designing and installing circuits so that the accumulation of protective currents and/or earth leakage currents downstream of the RCD does not exceed 30 % of the rated residual current of the device.
- Coordination of RCDs, general purpose, selective and time-delayed (circuit-breaker incorporating residual current protection (CBR) (Section 536 Selectivity refers).
- Coordination of RCDs with surge protective devices (SPDs) according to Regulation 534.4.7.

Where an RCD incorporates a time-delay the device should be marked with the symbol shown. One suitable device is an S-type RCD in accordance with *BS EN 61008-1* or *BS EN 61009-1*.

S

It should be noted that where selectivity between RCDs is required the details of the methods used to verify that selectivity has been achieved in accordance with Regulation 536.4.1.4 of *BS 7671*, should be appended to the Electrical Installation Certificate or Domestic Electrical Installation Certificate, as appropriate. The same applies where selectivity is used between any of the devices covered by Section 536.

Guidance on testing and verifying the operation of RCDs is provided in Chapter 8 'Testing'.

Confirmation overvoltage protection (SPDs) provided, where specified

Where surge protective devices (SPDs) have been specified for protection against overvoltage, the relevant requirements of Section 534 should be confirmed. Section 534 is applicable to AC power circuits but the requirements may also be followed for DC power circuits.

In particular, the type and location of an SPD should comply with the relevant requirements of Regulation Group 534.4.1, and in the event failure of an SPD fault protection must remain effective (Regulation 534.4.6 refers).

Where an SPD is installed downstream of a distribution board, identification of its presence, typically a label, should be provided at the board. An SPD installed on the load side of an RCD should have an immunity to surge currents of at least 3 kA (8/20) (Regulation 534.4.7 refers)..

Note: Installation of Type 1 SPDs downstream of an RCD is not recommended

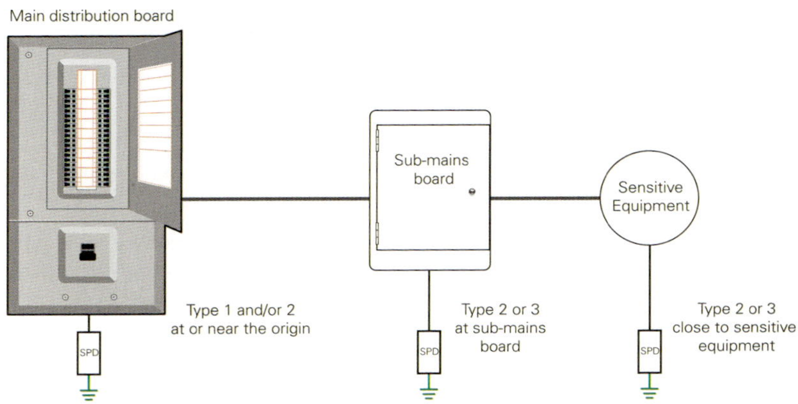

Fig 6.7.7 Example of the installation of Type 1, Type 2 and Type 3 SPDs (Regulation 534.4.1.5 refers)

To minimise the effects of inductance, conductors and interconnections associated with the connection of SPDs should be kept as straight and as short as possible (Regulation 534.4.8 refers). For a Type 1 SPD installed at or near the origin, the conductors installed between the SPD and the MET or the protective conductor should have a cross-sectional area (csa) of at least 16 mm^2 copper or equivalent. For a Type 2 SPD the conductors should have a csa of at least 6 mm^2 copper or equivalent (Regulation 534.4.10 refers).

Indication of SPDs continued functionality confirmed
SPDs should be provided with an indication status, in accordance with their product standard, so that it can be readily identified when the device no longer provides protection against overvoltage. Correct indication should be verified.

Selection of protective device(s) and base(s): correct type and rating
The rated current, settings (where applicable) and characteristics of the protective devices installed must be checked to verify that they are in accordance with the design information for the installation.

Fuses/circuit-breakers should be selected so that they are compatible with the particular circuit characteristics such as, for example, the starting currents of inductive loads.

The arrangement of fuse bases should prevent the possibility of contact between two adjacent bases and the fuse carrier (Regulation 533.1.2.1 refers). Fuse-links likely to be removed or replaced by ordinary persons should satisfy the requirements of Regulation 533.1.2.2.

A circuit-breaker intended for use by ordinary persons must be designed so the setting or calibration of its overcurrent release cannot be modified without the use of key or tool (Regulation 533.1.3 refers).

Single-pole protective devices in line conductors only
Inspection should confirm that all single-pole switching and control devices are connected in the line conductor only (Regulation 530.3.3 refers). No single-pole switching device, circuit-breaker, fuse or solid link which can be removed without the use of a tool or key or the like should interrupt or control the neutral. This check is made in addition to the polarity testing described in Chapter 8, of this Guide.

Protection against mechanical damage where cables enter equipment
Cables entering metallic enclosures should be suitably protected against strain and abrasion as required by Regulation 522.8.5 and 522.8.11 respectively. Inspection should confirm that, where necessary, suitable bushes and glands have been installed and are fitted in accordance with manufacturer's instructions.

Protection against electromagnetic effects where cables enter ferromagnetic enclosures
Single cables of AC circuits run in steel conduit or any other form of steel enclosure should be installed so that all associated line, neutral and, where provided, separate protective conductors are installed within the same enclosure (Regulation 521.5.1 refers). This requirement also applies to cables passing through holes in ferromagnetic enclosures (such as metallic distribution boards/ consumer units, steel trunking and the like). For example, the conductors of a three-phase circuit should be routed through the same enclosure rather than being routed through separate conduits, as shown in Fig 6.7.8.

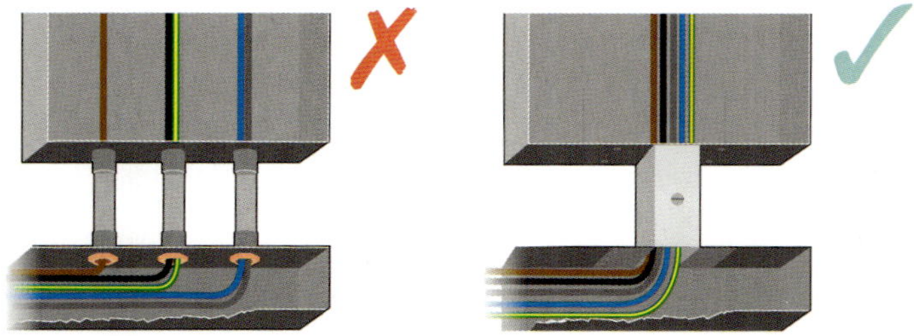

Fig 6.7.8 Protection against electromagnetic effects where cables enter ferromagnetic enclosures

Note: The use of single-core cables armoured with steel wire or tape is not permitted for AC circuits. However, the use of aluminium armour may be considered.

Confirmation that ALL conductor connections, including connections to busbars are correctly located in terminals and are tight and secure

The consequences of loose or poorly terminated conductors can cause high resistance connections leading to arcing and in some cases fire. Therefore, the electrical and mechanical integrity of connections in consumer units/distribution boards, panels and similar equipment should be checked and their tightness confirmed. (See also 'Termination of cables at enclosures' in Section 8 of this Chapter).

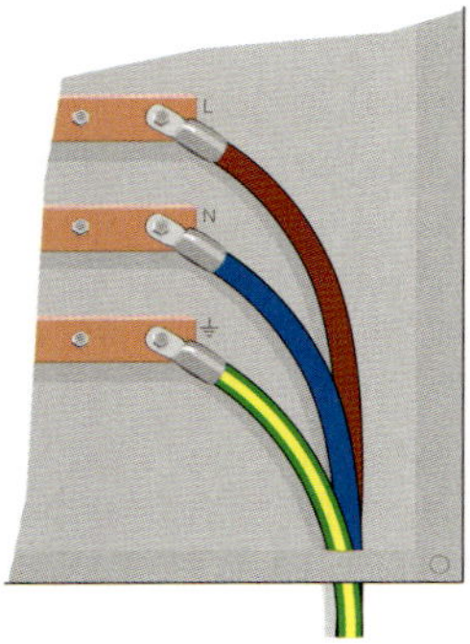

Note: The insulation and / or sheath of a cable connected to a busbar must be suitable for the maximum operating temperature of the busbar (Regulation 526.4 refers). Information on busbar and powertrack systems is contained in Appendix 8 of *BS 7671*

Fig 6.7.9 Confirmation that conductor connections are correctly located in terminals and are tight and secure

The following items on the schedule require the inspector to confirm that information, warning notices and labels have been provided as required by *BS 7671*.

7.21	Presence of RCD six-monthly test notice, where required:	(.........)
7.22	Presence of diagrams, charts or schedules at or near each distribution board, where required:	(.........)
7.23	Presence of next inspection recommendation label:	(.........)
7.24	Presence of non-standard (mixed) cable colour warning notice at or near the appropriate distribution board, where required:	(.........)
7.25	Presence of other required labelling:	(.........)

Fig 6.7.10 Distribution equipment

Presence of RCD six-monthly test notice, where required

Where, as will generally be the case, the installation incorporates one or more RCDs, it should be confirmed that a notice(s), such as the one shown in Fig 6.7.11, is provided at or near each RCD (Regulation 514.12.2 refers).

This installation, or part of it, is protected by a device which automatically switches off the supply if an earth fault develops. Six-monthly by pressing the button marked 'T' or 'Test'. The device should switch off the supply and should then be switched on to restore the supply. If the device does not switch off the supply when the button is pressed, seek expert advice.

Fig 6.7.11 RCD test notice

Note: The presence of a notice at (or near) the origin should be confirmed for an installation which incorporates one or more arc fault detection devices installed in accordance with Regulation 532.6.

Presence of circuit charts or schedules

As required by Regulation 514.9.1, the presence of the necessary diagrams, charts or tables should be confirmed. The form of information should be legible and durable, and should indicate all of the following.

- The type and composition of each circuit (that is, the points of utilisation served, number and size of conductors, type of wiring).
- The method used for compliance with Regulation 410.3.2 (that is, the protective measures used for basic and fault protection).
- The information necessary for the identification of each device performing the functions of protection, isolation and switching, and its location.
- Circuit(s) or equipment vulnerable to electrical testing.

For a typical domestic premises and other simple types of installations, providing a copy of the combined 'Schedule of Circuit Details and Test Results', which forms part of an electrical certificate supplied by NICEIC or ELECSA, fixed within or adjacent to the consumer unit is likely to achieve compliance with Regulation 514.9.1. For more complex installations, more comprehensive information will be required.

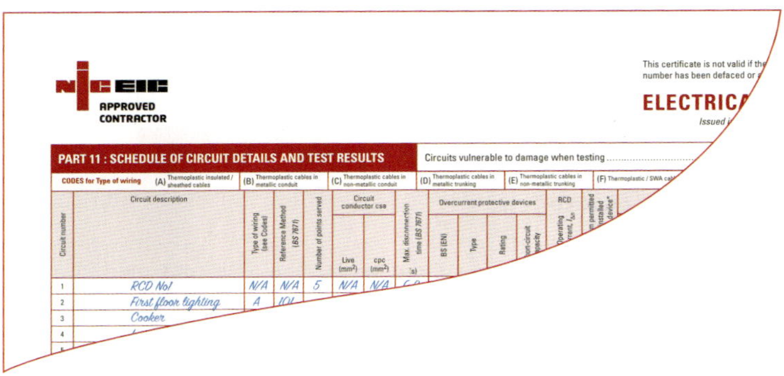

Fig 6.7.12 Schedule of Circuit Details and Test Results

Presence of next inspection recommendation label

A label should be fitted on completion of installation work (including additions and alterations) and on each completion of periodic inspection, in a prominent position at or near the origin, giving the recommended date within which the installation should be re-inspected (Regulation 514.12.1 refers).

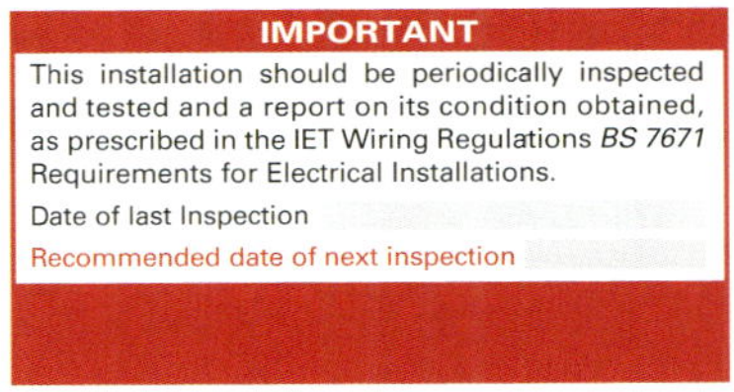

Fig 6.7.13 Next inspection recommendation label

Presence of non-standard (mixed) cable colour warning notice

Where as a result of wiring additions, alterations, extensions or repairs made to an installation, the installation contains wiring that has colours to different (older) versions of *BS 7671*, a warning notice must be fixed at or near the appropriate distribution board or consumer unit (Regulation 514.14.1 refers). (Refer to 'Circuits - Identification of conductors'.)

Fig 6.7.14 Presence of non-standard (mixed) cable colour warning notice

Presence of other required labelling

The inspector should verify that all relevant warning notices and labels have been fitted in the appropriate locations, in accordance with Section 514 of *BS 7671*. Common types include:

- **Alternative supplies:** (see Section 2 - 'Parallel or switched alternative sources of supply')
- **Earthing and bonding labels:** (see Section 3 - 'Automatic disconnection of supply')
- **A voltage warning:** The warning should be visible before gaining access to live parts in an enclosure where a nominal voltage exceeding 230 V to earth exists but would not normally be expected (Regulation 514.10.1 refers)

- **High protective conductor current:** Where a circuit is likely to have a high protective conductor current, information should be provided at the distribution board or consumer unit indicating the circuit(s).The information should be positioned so it is visible to a person modifying or extending the circuit (Regulation 543.7.1.205 refers).

Each protective device should be arranged and identified so that the circuit protected may be easily recognised (Regulation 514.8.1 refers). Except where there is no possibility of confusion, each item of apparatus for operation, regulation or other control should be labelled or have a suitable means of identification to indicate its purpose and the installation, or circuit, it controls (Regulation 514.1.1 refers).

Fig 6.7.15 Labelling of protective devices, switchgear and controlgear

6.8 Circuits

The items listed under this section of the 'Schedule of Items Inspected' relate to the installed circuits. The inspections carried out should be appropriate for the installation.

8. Circuits		
8.1	Identification of conductors:	(.........)
8.2	Cables correctly supported throughout, with protection against abrasion:	(.........)
8.3	Examination of cables for signs of mechanical damage during installation:	(.........)
8.4	Examination of installation of live parts, not damaged during erection:	(.........)
8.5	Non-sheathed cables protected by enclosure in conduit, ducting or trunking:	(.........)
8.6	Suitability of containment systems (including flexible conduit):	(.........)
8.7	Correct temperature rating of cable insulation:	(.........)

Fig 6.8.1 Inspection of circuits

Identification of conductors

Inspection should confirm that the relevant requirements of Section 514 of *BS 7671* have been satisfied. The conductors of cables, and of any bare conductors, should be correctly identified in terms of their polarity or function, and the wiring should be arranged and / or marked so that it is readily identified for the purposes of inspection, testing, repair or alteration (Regulation 514.1.2 refers).

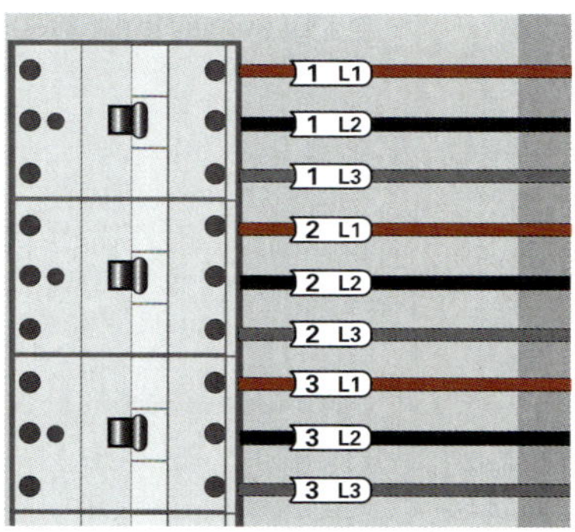

Fig 6.8.2 Wiring arrangement at a three-phase distribution board

Except where identification is not required, the cores of cables should be identified by colour and / or lettering, and / or numbering (Regulations 514.3.1 refers). Table 51 of *BS 7671* provides the colour and identification systems that should be used.

Note: A PEN conductor should be identified in accordance with Regulation 514.4.3.

The colour combination of green and yellow must be used for the identification of protective conductors only (Regulation 514.4.2 refers). The single colour green should not be used for the identification of protective conductors, functional earthing and bonding conductors and live conductors in power circuits (Regulation 514.4.5 refers). Where a circuit includes a neutral (or midpoint) conductor which is identified by colour, the colour blue must be used, and where a number is used to identify a neutral or midpoint a zero '0' should be used (Regulation 514.4.1 and Regulation 514.5.4 refer).

Colours to previous versions of *BS 7671*

Except where there is no possibility of confusion, unambiguous marking should be provided at the interface between conductors identified in accordance with *BS 7671* and wiring carried out to previous versions of the Regulations using the older (non-harmonised) colours. Guidance on identification at the interface between such conductors is provided in Appendix 7 of *BS 7671*.

As shown in Fig 6.8.3, where conductors having harmonised and non-harmonised colour identification are present in the same installation a 'Caution' notice in accordance with Regulation 514.14.1 should be provided at the appropriate consumer unit or distribution board.

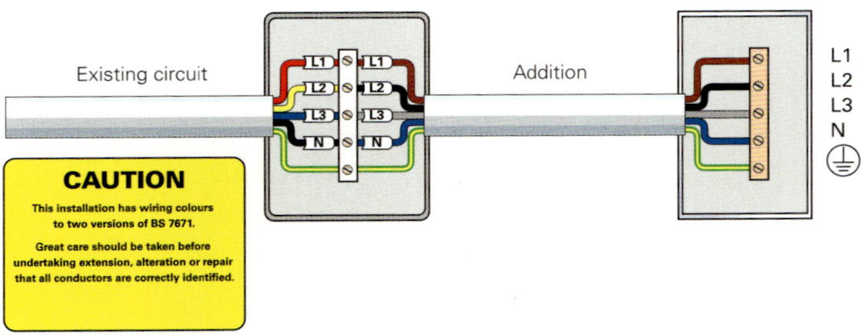

Fig 6.8.3 Typical arrangement for colour identification at an interface of different colour coded conductors

Note: Identification is not required for concentric conductors of cables, metal sheath or armour of cables when used as a protective conductor and for the other exceptions listed in Regulation 514.6.1.

Cables correctly supported throughout, with protection against abrasion

As required by Regulation 522.8.5, supports must be sufficient in number to prevent cables (including consumer unit meter tails) from being damaged by their own weight or by electromechanical forces associated with fault conditions. Cables should be adequately protected against abrasion and connection to enclosures should be made using the appropriate gland/termination in accordance with the cable manufacturer's instructions. Inspection should confirm that no appreciable strain is placed on conductor terminations.

Inspection should also confirm that a sufficient number of non-combustible (metal) supports have been installed to prevent, as shown in Fig 6.8.4, cabling from collapsing prematurely in the event of fire and subsequently obstructing routes (Regulation 521.10.202 refers). It should be noted that the extent of Regulation 521.10.202 is not restricted to any particular cable type or voltage, but applies to all wiring systems within the scope of *BS 7671*. The requirements apply not only to escape routes but wherever there is a risk of cables collasping prematurely in the event of fire and causing an obstruction within premises. Where there is a risk of premature collaspe of cables, inspection should confirm that the requirements of Regulation 521.10.202 are satisfied.

Cables installed in or on steel containment systems, such as metallic conduit, trunking or tray are deemed to satisfy Regulation 521.10.202.

Fig 6.8.4 Non-combustible supports should be used to prevent premature cable collapse

Examination of cables for signs of mechanical damage during installation

Generally, the risk of mechanical damage to cables is greater during construction than when the installation is in service. Therefore, at appropriate times during construction, such as prior to being enclosed, cables should be inspected for signs of damage and to verify that they are not at risk of damage due to the manner of their installation.

For example, where cables, conduits or ducts are buried in the ground, inspection is required to verify that the cables are at sufficient depth to avoid being damaged by any likely disturbance of the ground (Regulation 522.8.10 refers).

Where possible underground cables should be routed through areas of low risk. Where cables are routed through areas of higher risk of mechanical damage such as those used for planting, the cable depth should be increased appropriately and / or additional mechanical protection be provided to protect the cable from damage.

Cable sheaths that provide mechanical protection should extend into enclosures and bends formed in cables should not be less than the minimum recommended for the particular type and overall diameter of the cable (Regulation 522.8.3 refers).

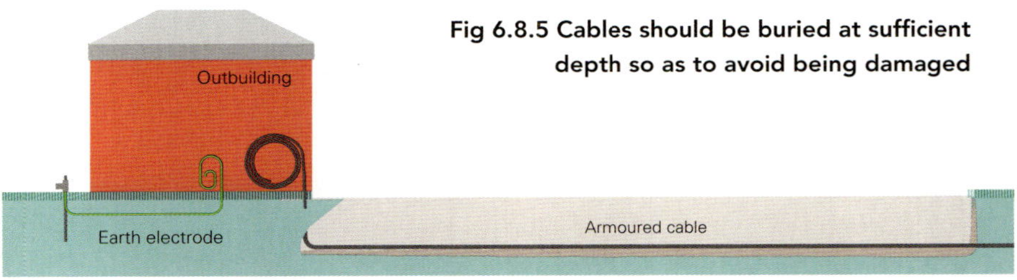

Fig 6.8.5 Cables should be buried at sufficient depth so as to avoid being damaged

Examination of insulation of live parts, not damaged during erection

Prior to the installation of cables or conductors, conduit, trunking and the like should be checked for burrs and sharp edges. Inspection should also confirm that a sufficient number of access boxes have been provided to facilitate the drawing in of cables and conductors (Regulation 522.8.6 refers).

Holes or slots that have been cut in trunking or enclosures to allow conductors to pass through should have suitable protection against abrasion, such as insertion of an appropriate bush or grommet strip (Regulation 522.6.1 refers).

Non-sheathed cables protected by enclosure in conduit, ducting or trunking

Except for a protective conductor complying with Section 543, non-sheathed cables should be enclosed in conduit, ducting or trunking (Regulation 521.10.1 refers).

After the cables have been installed, it should be confirmed that all the appropriate lids/covers have been fitted, in accordance with the manufacturer's instructions.

Suitability of containment systems (including flexible conduit)

Inspection should confirm that trunking conduit and other types of containment are adequately protected against the external influences likely to be encountered, and the components used to connect conduit, trunking and enclosures together, such as couplers, bushes and locknuts are suitable and correctly installed. Details of external influences are given in Section 522 of *BS 7671*.

Containment should comply with the appropriate product standard. Regulation 521.6 specifies the following harmonised product standards for containment:

- Conduits should comply with the appropriate part of the *BS EN 61386* series.
- Trunking or ducting should comply with the appropriate part of the *BS EN 50085* series.
- Tray and ladder systems should comply with *BS EN 61537*.

Containment installed in an environment prone to damp / wetness should be adequately protected against corrosion. For example, where a metallic conduit is installed outdoors inspection should confirm that the conduit is a galvanised type and has no exposed threads at connection boxes and enclosures (Fig 6.8.6).

Where necessary to prevent the ingress of water, conduits should be fitted with appropriate seals or gaskets, in accordance with the conduit manufacturer's instructions, and any protective coatings/coverings removed during installation (say, as a result of threading or cutting the conduit) should be properly reinstated, such as by applying a galvanised paint.

Allowance for the expansion of containment systems should be provided where necessary. For example, where PVC conduit or trunking is installed in locations exposed to high ambient temperatures, expansion couplers should be installed at appropriate intervals according to the manufacturer's instructions.

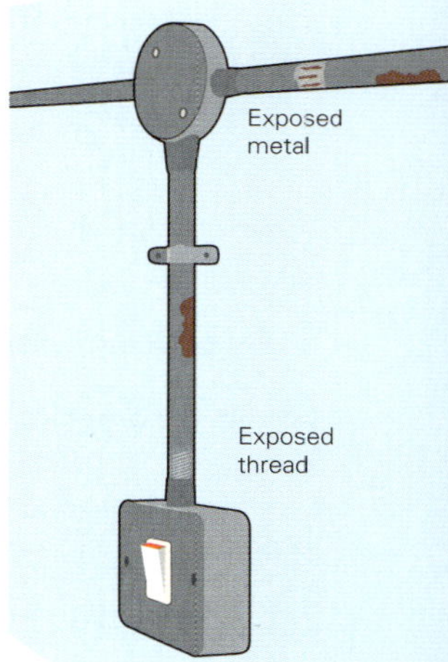

Fig 6.8.6 Conduit exposed to corrosion

A flexible conduit should be used for the connection of an item of equipment that is subject to vibration in normal use, such as a motor, and irrespective of whether the flexible conduit is of metallic construction or not, it must not be used to perform the function of a protective conductor (Regulation 543.2.3 refers).

Note: A flexible conduit, of metallic construction or otherwise, is not suitable to meet the mechanical protection requirements of Regulation 522.6.204 regarding cables concealed in walls or partitions (refer to: 'Cables installed under floors above ceilings, in walls / partitions, adequately protected against damage').

8.8	Adequacy of cables for current-carrying capacity with regard to the type and nature of installation:	(.........)
8.9	Adequacy of protective devices: type and fault current rating for fault protection:	(.........)
8.10	Adequacy of AFDD(s), where specified:	(.........)
8.11	Presence and adequacy of circuit protective conductors:	(.........)
8.12	Coordination between conductors and overload protective devices:	(.........)
8.13	Wiring systems and cable installation methods / practices appropriate to the type and nature of installation and external influences:	(.........)
8.14	Cables concealed under floors, above ceilings, in walls / partitions, adequately protected against damage:	(.........)
8.15	Cables installed in walls / partitions, installed in prescribed zones:	(.........)
8.16	Provision of additional protection by RCDs having rated residual operating current ($I_{\Delta n}$) not exceeding 30 mA:	
	a) For all socket-outlets with a rated current not exceeding 32 A or less, unless exempt	(.........)
	b) For supplies to mobile equipment with a current rating not exceeding 32 A for use outdoors	(.........)
	c) For cables concealed in walls / partitions at a depth of less than 50 mm	(.........)
	d) For cables concealed in walls / partitions containing metal parts regardless of depth	(.........)
	e) For circuits supplying luminaires within domestic (household) premises only	(.........)

Fig 6.8.7 Inspection of circuits

Correct temperature rating of cable insulation

The current to be carried by any conductor for sustained periods during normal operation (including any harmonic current) should be such that the temperature limit specified in Table 52.1 of *BS 7671* is not exceeded (Regulation 523.1 refers).

Cables selected in accordance with the current carrying capacities given in Appendix 4 of *BS 7671*, subject to any necessary rating factors, are considered to satisfy the requirements of Regulation 523.1.

Adequacy of cables for current-carrying capacity with regard to the type and nature of installation

The sizes and types of all the installed cables should be checked against the installation design information, to confirm that they have been correctly selected and installed to meet the requirements of Sections 523, 524 and 525 of *BS 7671* with regard to selection of conductors for current-carrying capacity and voltage drop respectively. Tables of current-carrying capacity are given in Appendix 4 of *BS 7671*.

The cross-sectional area of each conductor should not be less than the minimum values specified in Table 52.3 of *BS 7671*. Where a cable is totally surrounded by thermal insulation its current-carrying capacity should be reduced appropriately in accordance with Table 52.2 of *BS 7671* (Regulation 523.9 refers).

Note: Power, control and communications cables should comply with the Construction Products Regulation (CPR) (see Appendix 2, item 17, of *BS 7671*).

Adequacy of protective devices: type and fault current rating for fault protection

Inspection should verify that devices used to provide fault protection, such as fuses, circuit-breakers, RCDs and the like, comply with Regulation 533.1.1 and satisfy the requirements of Regulation Group 434.5.

As required by Regulation 434.5.1, it should be verified that the rated short-circuit breaking capacity of each device providing fault protection is not less than the maximum prospective fault current at the point at which the device is installed. Where this is not the case, it must be confirmed that fault protection is provided by another device (back-up device) installed on the supply side.

For single-phase installations the prospective fault current is the greater of either the fault current between the line conductor and neutral or the fault current between line conductor and Earth.

Fig 6.8.8 Fault protection provided by the 63 A device installed on the supply side of the protective devices.

Note: Where a consumer unit to *BS EN 61439-3* is installed in a domestic (household) installation it has a conditional rating of 16 kA.

RCDs should be inspected to confirm that their rated residual operating currents are suitable to provide fault protection, and the effectiveness of any test facility forming part of an RCD should be verified (Regulation 643.10 refers). Where an RCD is used to provide fault protection an overcurrent protective device must also be provided (Regulation 411.4.5 refers).

Note: In certain circumstances, such as where disconnection of the circuit may result in danger, devices for protection against fault current may be omitted (Regulation 434.3 refers).

Arc fault detection devices provided for protection against fire, where specified

Where the installation design specifies one or more arc fault detection devices (AFDDs) to provide additional protection against fire, inspection should confirm that these devices comply with *BS EN 62606*, and are installed in AC final circuits not exceeding 230 V. Each AFDD should protect one final circuit only and be installed at the origin of the circuit it is intended to protect (Regulation 532.6 refers).

Note: An AFDD is an additional form of protection against the risk of fire and must be used in conjunction with an appropriate overcurrent protective device.

Fig 6.8.9 Arc fault detection device integrated with a protective device

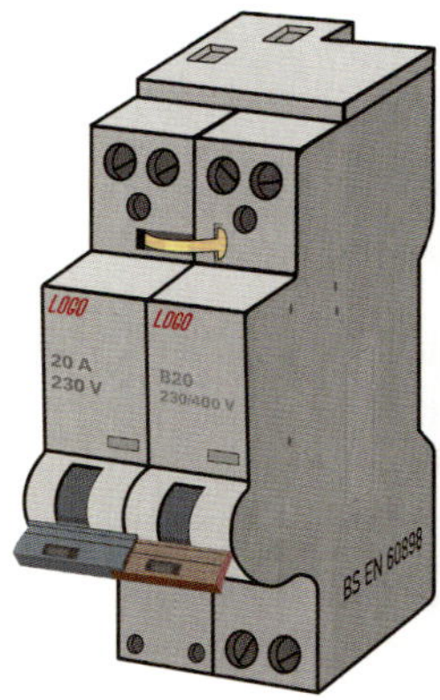

Presence and adequacy of circuit protective conductors

A visual inspection must be made to check that circuit protective conductors, where required, have been correctly selected and installed. The designer may have determined the cross-sectional area of the circuit protective conductors either by calculation or by application of Table 54.7 of *BS 7671*.

Unless installed in a conduit or duct that provides an equivalent level of protection against mechanical damage (or supplied from a SELV source), a cable buried in the ground should incorporate earthed armour or metal sheath, or both, that is suitable for use as a protective conductor (Regulation 522.8.10 refers). Elements that may affect the continuity of the connection should be checked; the tightness of the gland connection, the presence of paint, corrosion and the like.

Coordination between conductors and overload protective devices

Confirmation is required that each protective device providing overload protection satisfies the conditions specified in Regulation 433.1.1. Where the device is a semi-enclosed fuse to *BS 3036*, Regulation 433.1.202 should be taken into account. Where necessary the selection of protective devices should take into account the presence of harmonic currents (Regulation 533.2.2 refers).

Wiring systems and cable installation methods / practices appropriate to the type and nature of installation and external influences

Cables and wiring systems and their method of installation must be suitable for the external influences likely to be present (Section 522 refers).

For example, a cable installed in a space where thermal insulation is likely to be applied, such as a loft, should wherever practical be fixed so that it is not covered by thermal insulation (Regulation 523.9 refers). Cables and wiring systems installed in escape routes of buildings should comply with the requirements of Regulation 422.2.1.

A wiring system installed in a location where combustible or flammable materials, such as wood, paper or textiles, are used or stored as part of the normal work activities, should be selected and erected so that the risk of fire is minimised. Therefore, inspection should confirm that wiring systems installed in such locations satisfy the additional requirements of Regulation Group 422.3 'Locations with risks of fire due to the nature of processed or stored materials'.

Note: A classification of external influences is given in Appendix 5 of *BS 7671*.

Fig 6.8.10 A location with risks of fire due to the nature of processed or stored materials

Cables concealed under floors, above ceilings, in walls/partitions, adequately protected against damage

To confirm compliance with Regulation 522.6.201, inspection should be undertaken at the appropriate time to verify that cables concealed under floors or above ceilings have been installed so that they are not liable to damage by contact with the floor, the ceiling or any of their fixings.

Where a cable has to pass through a timber joist within a floor or ceiling construction, or through a ceiling support, the cable should:

1. be at least 50 mm, measured vertically from the top, or bottom as appropriate, of the joist or batten, or
2. comply with Regulation 522.6.204.

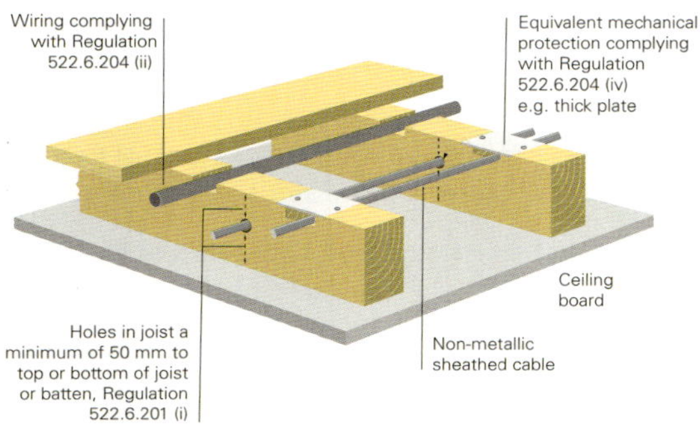

The position and size of holes and notches in joists etc. is subject to the requirement of the Building Regulations.

Fig 6.8.11 Cables in floor joists

Cables in walls/partitions installed in prescribed zones

Cables installed in walls or partitions can be vulnerable to damage from nails, screws and the like, which can lead to the danger of electric shock or fire. Therefore, inspection should be undertaken at the appropriate stage to confirm that cables installed in walls or partitions at a depth of less than 50 mm from the surface are installed within the prescribed zones specified by Regulation 522.6.202, or satisfy the requirements of Regulation 522.6.204.

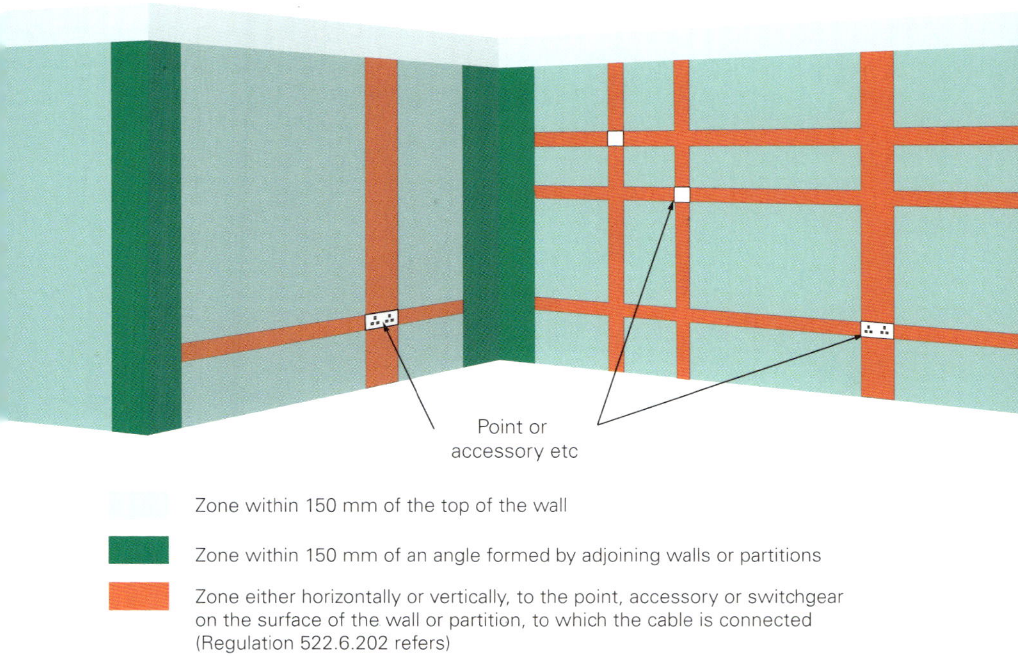

Fig 6.8.12 Prescribed zones of Regulation 522.6.202

Generally, a zone created on one side of a wall or partition does not extend to the reverse side. However, where the location of the accessory, point or switchgear can be determined from the reverse side, as shown in Fig 6.8.13, then a zone formed on one side of a wall or partition of 100 mm thickness or less extends to the reverse side.

The presence of the doorway in Fig 6.8.13, means that a person intending to drive a nail, screw or similar into the wall can reasonably determine the location of the accessory from the reverse side of the wall. A mirror image of the permitted cross-shaped zone created in the room with the accessory (Room 1) is therefore considered to exist in Room 2.

Note: Where additional mechanical protection is used to satisfy the requirements of 522.6.204, the protection provided should be sufficient (at least 3 mm thick) to prevent penetration of a cable by nails, screws and the like, except where shot-fired nails or similar fixings are used. The use of standard metallic cable capping is insufficient to prevent such penetration.

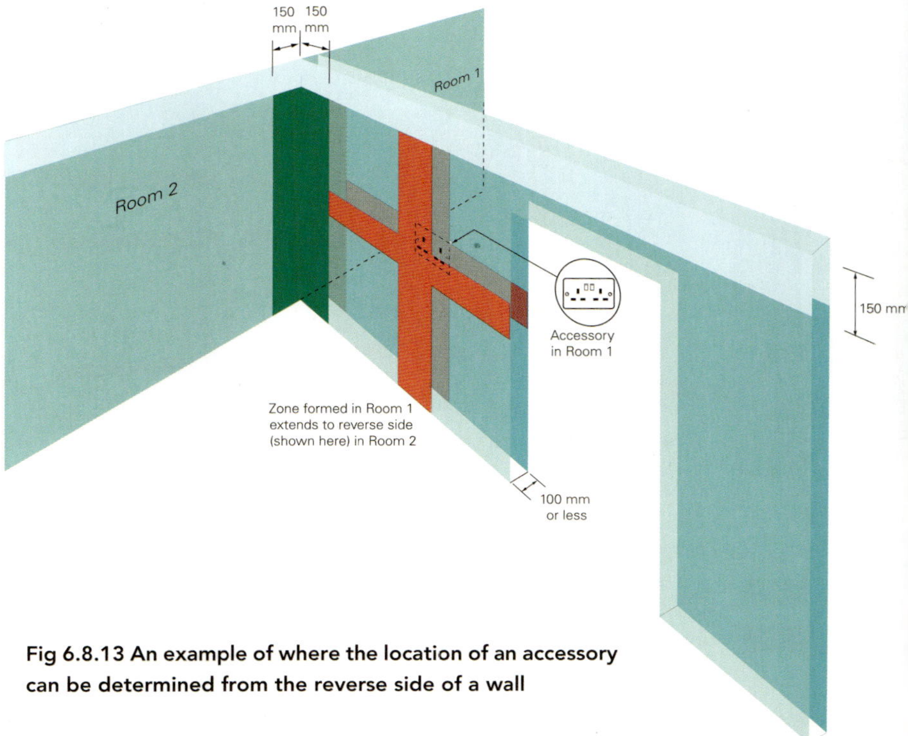

Fig 6.8.13 An example of where the location of an accessory can be determined from the reverse side of a wall

Provision of additional protection by RCDs having rated residual operating current ($I_{\Delta n}$) not exceeding 30 mA

Inspection should confirm that additional protection is provided in accordance with Regulation 415.1.1 by an RCD having a rated residual operating current ($I_{\Delta n}$) not exceeding 30 mA, for all the following:

- For all socket-outlets with a rated current not exceeding 32 A, unless exempt* (Regulation 411.3.3 refers).
- For supplies to mobile equipment with a rated current not exceeding 32 A for use outdoors (Regulation 411.3.3 refers).
- Cables concealed in walls or partitions at a depth of less than 50 mm, unless complying with 522.6.204 (Regulation 522.6.202 refers).
- Cables concealed in walls or partitions containing significant metal parts regardless of depth, unless complying with 522.6.204 (Regulation 522.6.203 refers).

- For circuits supplying luminaires within domestic (household) installations (Regulation 411.3.4 refers).

*An exception is permitted by Regulation 411.3.3 for premises other than a dwelling where a documented risk assessment determines that RCD protection is not necessary. Such a decision should only be taken in agreement with the client or the person responsible for the safety of the premises and should involve an electrically skilled person(s) (Appendix 2, item 11 of *BS 7671* refers). The risk assessment must be appended to the Electrical Installation Certificate.

Note: The requirements of Regulation 411.3.3 are not applicable to FELV or reduced low voltage systems.

Each RCD installed to provide additional protection should be inspected to confirm that it complies with the relevant product standard and the required rated residual operating current. Additionally, operation of any RCD should not result in unreasonable inconvenience (Regulation 314.1 refers). Verification of RCDs is covered in Chapter 8 'Testing'.

8.17	Provision of fire barriers, sealing arrangements so as to minimise the spread of fire:	(.........)
8.18	Band II cables segregated / separated from Band I cables:	(.........)
8.19	Cables segregated / separated from non-electrical services:	(.........)
8.20	Termination of cables at enclosures:	
	a) Connections under no undue strain	(.........)
	b) No basic insulation of a conductor visible outside enclosure	(.........)
	c) Connections of live conductors adequately enclosed	(.........)
	d) Adequately connected at point of entry to enclosure	(.........)
8.21	Suitability of circuit accessories for external influences:	(.........)
8.22	Circuit accessories not damaged during erection:	(.........)
8.23	Single-pole devices for switching or protection in line conductors only:	(.........)
8.24	Adequacy of connections, including cpcs, within accessories and at fixed and stationary equipment:	(.........)

Fig 6.8.14 Circuits

Provision of fire barriers, sealing arrangements so as to minimise the spread of fire

Where wiring systems pass through building elements of a specified fire resistance, such as walls or floors, inspection is required to verify that both external and internal sealing, having a fire resistance no less than that prescribed for the building element, before penetration, is provided (Regulation 527.2.1 refers).

Internal sealing is not required for a non-flame-propagating wiring system with an internal cross-sectional area of not more than 710 mm^2 that satisfies the requirements of Regulation 527.2.3.

For new installations, the presence of the required fire barriers should be verified at the appropriate stage, when access to the wiring system penetration is readily available.

Fig 6.8.15 Internal and external fire-sealing where trunking passes through a wall

Note: An internal cross-sectional area of 710 mm^2 is equivalent to a 32 mm diameter conduit.

Band II cables segregated/ separated from Band I cables

Separation or segregation must be provided between Band I and Band II circuits. Compliance may be confirmed where every cable or conductor is insulated for the highest voltage present, or where Band I and Band II circuits are installed in separate containment or by using one of the other methods listed in Regulation 528.1.

Note: For fire alarm and emergency lighting circuits, segregation should be undertaken in accordance with *BS 5839* series and *BS 5266* series respectively.

Where systems operating at different voltage bands and frequencies are installed in close proximity, such as power, communications, signalling and data cables, there is an increased risk of electromagnetic interference. Therefore, requirements and recommendations intended to minimise the risk of electromagnetic disturbances due to the switching and operation of certain inductive loads, such as variable speed drives, transformers and fluorescent luminaires, are contained in Section 444 of *BS 7671*.

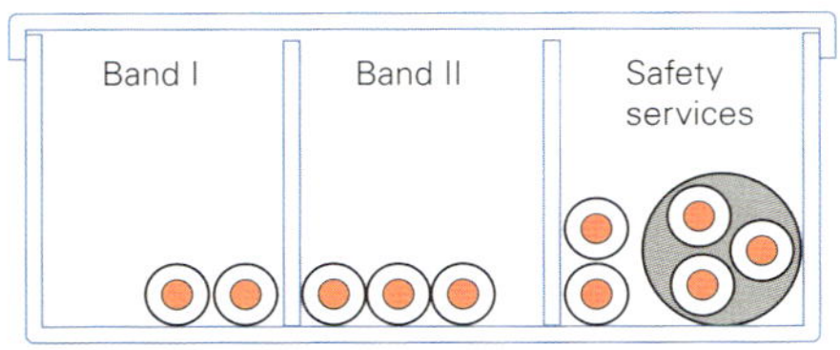

Fig 6.8.16 The use of compartmental trunking to provide separation between different Voltage Bands

Note: Unless the methods specified in Regulation 528.2 are applied, a minimum clearance of 100 mm should be maintained between underground power and communications cables.

Cables segregated/separated from non-electrical services

Inspection is required to verify that wiring systems are routed and positioned, or shielded, so that they will not suffer harmful effects from any heat, steam or condensation produced by non-electrical services (Regulation 528.3.1 refers). Where a wiring system is installed near one or more non-electrical services, sufficient space or shielding should be provided to minimise the risk of damage from any foreseeable maintenance work (Regulation 528.3.3 refers).

Note: Where there is a lift, a check should be made to verify that the lift or hoist well has not been used as a route for cables which do not form part of the lift installation, as defined in *BS EN 81* series.

Termination of cables at enclosures

Connections between conductors and between conductors and equipment should be inspected to confirm that they are sound, both mechanically and electrically (Regulation 526.1 refers). In addition to confirming that there is no appreciable mechanical strain on the terminations and no insulation visible outside enclosure inspection should verify that:

- **Connections of live conductors are adequately enclosed:** Every joint or connection in a live conductor or PEN conductor must be made in an enclosure complying with Regulation 526.5. This requirement applies not only to connections in low voltage circuits, but also those in extra-low voltage circuits. The enclosure must also provide adequate protection against mechanical damage, stress and external influences that it is likely to be subjected.
- **Adequate connection at point of entry to enclosure:** Where a cable is connected to an enclosure using a gland or bush, inspection should confirm that the correct type of gland or bush has been installed to prevent strain (in particular, armoured cable glands must not be used as a means of support for cables). The method of connection should take account of external influences such as moisture, vibration and thermal cycling.

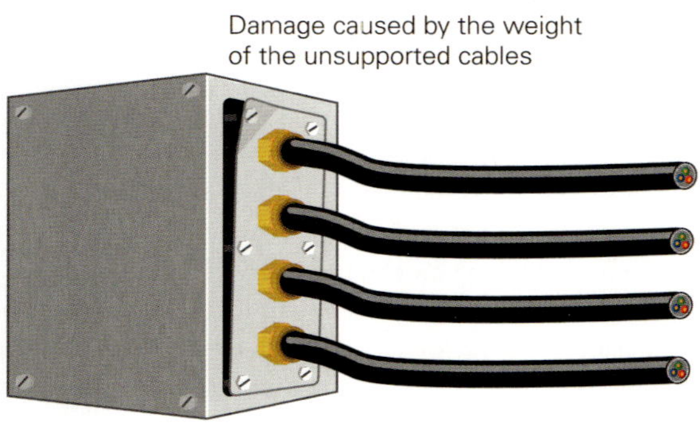

Fig 6.8.17 Damage to enclosure caused by unsupported SWA cables

Suitability of circuit accessories for external influences

Accessories should be suitable (comply with the appropriate British or Harmonised Standard and the required IP rating) for the external influences likely to be encountered (Regulation 512.2.1 refers). Where an accessory is located in an enclosure or cabinet to protect it against external influences, it should not impair its operation or cause overheating or impede inspection, testing and maintenance of the protected accessory. Classification of external influences is contained in Appendix 5 of *BS 7671*.

Circuit accessories not damaged during erection

Installed accessories should be inspected to verify that have not been damaged during the construction process.

Single-pole devices for switching or protection in line conductors only

Inspection is required to verify that all single-pole switching and protective devices are connected in the line conductor only (Regulation 530.3.3 refers). Guidance on the selection of switching devices is provided in Table 537.4.

Note: No switching (or isolating) device should be inserted in a PEN conductor.

Adequacy of connections, including cpcs, within accessories and at fixed and stationary equipment

It should be confirmed that conductors, including circuit protective conductors, at accessories and other items of equipment are connected to the correct terminals, and that the connections are both electrically and mechanically sound.

Continuity testing of protective conductors is covered in Chapter 8 'Testing'.

6.9 Isolation and switching

This is the general term used to cover four basic functions:

- isolation,
- switching off for mechanical maintenance,
- emergency switching and
- functional switching.

Inspection is required to confirm that the relevant requirements of Chapter 46 are satisfied according to the particular function, and that devices used for isolation or switching comply with the relevant requirements of Chapter 53 of *BS 7671* (Regulation 461.1 refers).

Note: Chapter 46 does not cover electrically-powered equipment within the scope of *BS EN 60204*.

All devices used for isolation and switching should be suitable for the currents they will have to carry and, where appropriate, make or break, under normal and fault conditions. It should be noted that a single device may perform more than one of the functions listed providing it satisfies all the requirements for each function concerned (Regulation 530.3.4 refers). Devices suitable to perform the functions of isolation and switching are given in Table 53.4 of *BS 7671*.

In accordance with Regulation 461.2, isolation or switching of the neutral conductor is not required for TN-C-S and TN-S systems where: the neutral conductor is reliably connected to Earth and protective equipotential bonding is installed.

Note: In TN-C systems (and the TN-C part of TN-C-S systems) the PEN should not be isolated or switched (Regulation 461.2 refers).

9. Isolation and switching

9.1 Isolators:

a)	Presence and location of appropriate devices	(.........)
b)	Capable of being secured in the OFF position	(.........)
c)	Correct operation verified (functional check)	(.........)
d	The installation, circuit or part thereof that will be isolated is clearly identified by location and / or durable marking	(.........)
e)	Warning notice posted in situations where live parts cannot be isolated by the operation of a single device	(.........)

Fig 6.9.1 Isolators

Isolators

The purpose of isolation is to enable persons to carry out work on, or in the vicinity of, parts that are normally live in service without risk of injury from electric shock from those parts. In larger installations, additional isolating devices will be specified for operational reasons, so that a particular section can be safely worked on without having to isolate the whole installation. (see Appendix A 'Isolation procedure'.)

The following inspections are included on the schedule:

- **Presence and location of appropriate devices:** Inspection is required to confirm that each electrical installation has a means of isolation from each supply (Regulation 462.1 refers), and that a means of isolation is provided for every circuit, as required by Regulation 462.2. Whilst a multi-pole switching device which disconnects all applicable poles of the relevant supply is preferred, single-pole devices are permitted, subject to the provisions of Regulation 461.2 being met (Regulation 537.2.6 refers). Where appropriate, a common means of isolation may be provided for a group of circuits.

  Where a link is inserted in a neutral conductor for isolating purposes, it should be accessible to skilled persons only and be connected so that it can only be removed by a tool (Regulation 537.2.8 refers).

  Note: A semiconductor device should not be used to provide isolation (Regulation 537.2.2 refers).

- **Capable of being secured in the OFF position:** Devices for isolation should preferably be located adjacent to or near the equipment and to prevent inadvertent closure such devices should be secured (Regulation 462.3 refers). For an off-load isolating device, it may also be achieved by interlocking the isolator with a load-breaking device (Regulation 537.2.5 refers).

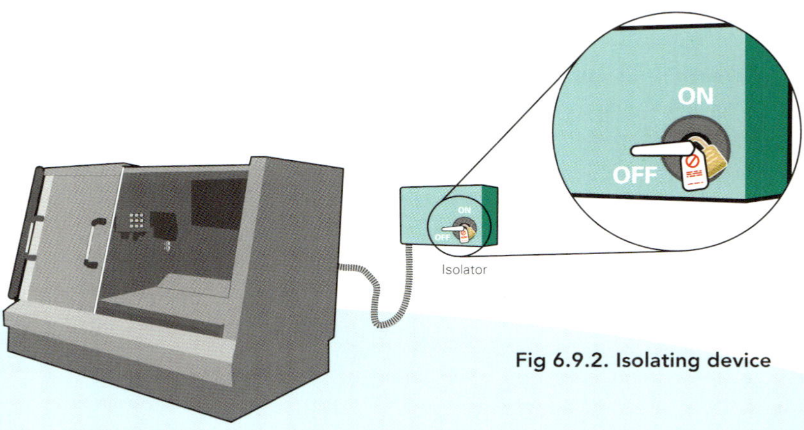

Fig 6.9.2. Isolating device

- **Correct operation verified (functional check):** A functional check should be carried out to check the operation of isolators, as required by Regulation 643.10.
- **The installation, circuit or part thereof that will be isolated clearly identified by location and/or durable marking:** Each isolating device should be clearly identified, either by its position or by durable marking, to indicate the circuit or installation it isolates (Regulation 537.2.7 refers).
- **Warning label notice posted in situations where live parts cannot be isolated by the operation of a single device:** Where live parts cannot be isolated by the operation of a single device, isolation is required at each position. Unless there is no possibility of confusion the location of each isolating device must be indicated (Regulation 514.11.1 refers).

Switching off for mechanical maintenance

Inspection of the following items is required to confirm that the selection and erection of devices intended to provide switching off for mechanical maintenance comply with the relevant requirements of Regulation Group 537.3.2.

9.2	Switching off for mechanical maintenance:		
	a)	Presence of appropriate devices	(.........)
	b)	Acceptable location (local or remote)	(.........)
	c)	Capable of being secured in the OFF position	(.........)
	d)	Correct operation verified (functional check)	(.........)
	e)	The installation, circuit or part thereof to be disconnected clearly identified by location and / or durable marking	(.........)

Fig 6.9.3 Switching off for mechanical maintenance

- **Presence of appropriate devices:** The purpose of switching off for mechanical maintenance is to enable persons to safely replace, refurbish or clean lamps, and to maintain non-electrical parts of electrical equipment, plant and machinery. For example, a means of switching off for mechanical maintenance would be required to enable maintenance personnel to clean luminaires having lamps that may cause burns, or to enable a mechanical fitter to work on the normally moving parts of an electrically-driven machine. Therefore, where mechanical maintenance may involve a risk of physical injury, inspection should verify that suitable devices for switching off have been provided (Regulation 464.1 refers).

Devices suitable for such purposes are given in Table 537.4 of *BS 7671* and include, for example:

- multi-pole switches,
- circuit-breakers,
- control switches that operate contactors, and
- plugs and socket-outlets.

A switch that is used to perform this function should be capable of cutting off the full load current to the relevant part of the installation (Regulation 537.3.2.2 refers).

Note: Where maintenance operations are likely to involve work on or near live parts, such as could be the case if the work involved the removal of covers housing live conductors, a means of isolation should be provided.

Fig 6.9.4 Switching off for mechanical maintenance

- **Acceptable location (local or remote):** Devices for switching off for mechanical maintenance must be suitably positioned for ready operation, and as such are generally positioned adjacent to the equipment requiring maintenance. Unless the device is to be continuously under the control of the person performing the maintenance, suitable provisions (for example, locking off facilities on the device) must be available to prevent the equipment from being reactivated during mechanical maintenance. Additionally, where the device is located remotely from the equipment, a notice should be fixed adjacent / near to the equipment stating where the device is located (Regulation 537.3.2.4 refers). In ticking this item on the schedule the inspector is confirming that devices located both locally and remotely comply with *BS 7671*.

- **Capable of being secured in the OFF position:** Where the switching device is under the supervision of the person undertaking the maintenance it would not necessarily need to be capable of being secured in the 'off' position, but if the device is located remotely from the equipment it must be capable of being secured in the 'off' position.The device should be inserted in the main supply circuit (Regulation 537.3.2.2 refers) and provide a clear and reliable indication of the open position of the contacts, such as by using a device marked with symbols 'O' and 'I'. The device or the switch controlling the device should be manually operated (Regulation 537.3.2.3 refers).

- **Correct operation verified (functional check):** The operation of devices provided for switching off for mechanical maintenance should be verified, as required by Regulation 643.10.

- **The installation, circuit or part thereof to be disconnected clearly identified by location and/or durable marking:** Devices provided for the purposes of switching off circuits so that mechanical maintenance can be undertaken, should be clearly identified, either by their position or by durable marking, to indicate their use (Regulation 537.2.7 refers).

Emergency switching/stopping

9.3	Emergency switching / stopping:		
	a)	Presence of appropriate devices	(.........)
	b)	Readily accessible for operation where danger might occur	(.........)
	c)	Correct operation verified (functional check)	(.........)
	d)	The installation, circuit or part thereof to be disconnected clearly identified by location and / or durable marking	(.........)
	e)	Firefighter's switches present, where required:	(.........)

Fig 6.9.5 Emergency switching /stopping

The purpose of emergency switching/stopping is to remove unexpected dangers, as quickly as possible (Regulation 465.1 refers). For example, provision of emergency switching for a cutting machine in a factory enables the operation of the machine to be stopped rapidly should anyone ever became caught or trapped in the machine.

Where devices for emergency switching off are provided the following items should be inspected.

- **Presence and condition of appropriate devices:** Devices provided for emergency switching off should be selected and erected in accordance with Regulation Group 537.3.3 and Regulation 537.2 (Regulation 537.3.3.1 refers). As required by Regulation 465.3, emergency switching should be achieved by one single action that results in interruption of the appropriate supply, and may be achieved by a single device or a combination of devices activated by a single action such as, for example, a contactor used in conjunction with the emergency switching device (Regulation 537.3.3.3 refers).

  Such a device should, wherever practicable, be hand-operated (Regulation 537.3.3.4 refers) and should be capable of breaking the full-load current of the relevant part of the installation (Regulation 537.3.3.2 refers). Due account must be taken of any stalled motor conditions.

  It should be noted that a plug and socket-outlet should not be selected for use as an emergency switching device.

- **Readily accessible for operation where danger might occur:** Devices provided for emergency switching off should be readily accessible and convenient for use at places where danger might occur (Regulation 537.3.3.6 refers).

- **Correct operation verified (functional check):** As required by Regulation 643.10, devices used for emergency switching off should be checked to verify they operate correctly. Where the operation and means of re-energisation for a device used for emergency switching off is unlikely to be controlled by the same individual, a device capable of latching in the 'OFF' position should be used (Regulation 537.3.3.7 refers).

Fig 6.9.6 Identification of a device for emergency switching

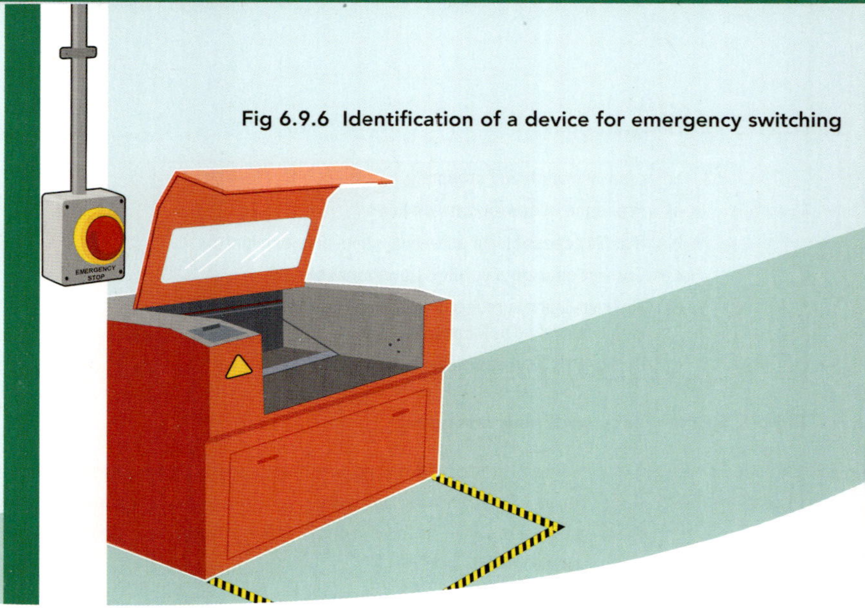

Note: The operation of a device used for emergency switching off should not be inhibited by any other operation carried out within the installation.

- **The installation, circuit or part thereof to be disconnected clearly identified by location and/or durable marking:** The operating means (for example a handle or push-button) must be clearly identifiable, preferably by colour. Where identification is provided by colour, red should be used with a contrasting background, such as yellow (Regulation 537.3.3.5 refers).

- **Firefighter's switches present, where required**: A firefighter's switch is an essential safety device that is provided to ensure that parts of an electrical installation designed to operate at a voltage higher than low voltage, such as outdoor lighting installations and indoor discharge lighting installations, can be readily isolated in an emergency.
  Inspection of these devices is required to confirm that each firefighter's switch has been selected and installed in accordance with the requirements of Regulation Group 537.4.

  A firefighter's switch complying with *BS EN 60669-2-6* or *BS EN 60947-3* (Regulation 537.4.1 refers), should be inserted in the low voltage circuit supplying the installation designed to operate at a voltage higher than low voltage (Regulation 537.4.2 refers). The switch should be easily visible, accessible and marked to indicate the installation or the part of the installation it controls (Regulation 537.4.3 refers).

The switch should be placed in a conspicuous position, reasonably accessible to firefighters, and mounted at a height not more than 2.75 m above the ground or the standing surface immediately below the switch (Regulation 537.4.2.2 refers). As required by Regulation 537.4.4, the following information should be durably marked on the switch using lettering having a height not less than 10 mm:

- The 'ON' and 'OFF' positions (the OFF (unpowered) position being at the top)
- 'FIREFIGHTER'S SWITCH' or 'FIRE SWITCH'

This information should be legible to a person standing on the ground at the intended site of the switch, without having to open the enclosure.

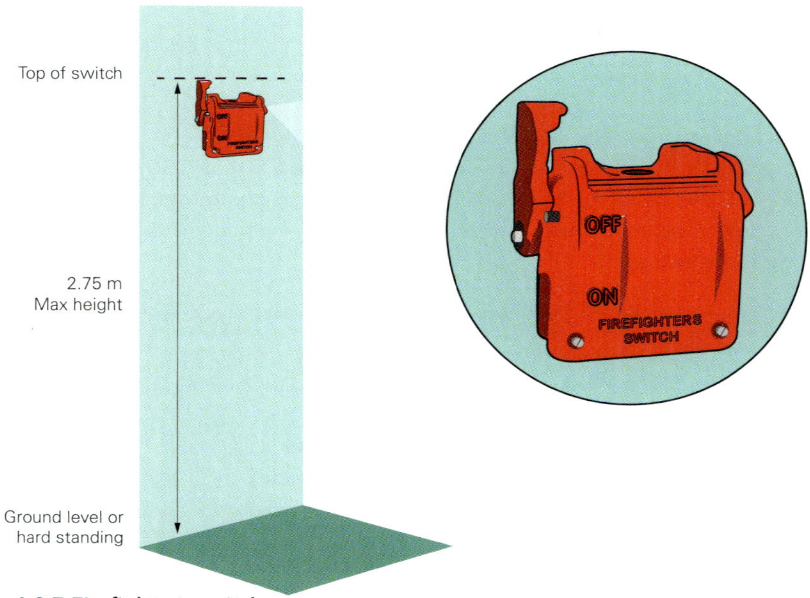

Fig 6.9.7 Firefighter's switch

Functional switching

Functional switching is required for any circuit or part of a circuit that may require to be controlled independently of other parts of the installation (Regulation 463.1.1 refers). The following items are included on the schedule for inspection.

9.4	Functional switching:	
	a) Presence of appropriate devices	(.........)
	b) Correct operation verified (functional check)	(.........)

Fig 6.9.8 Functional switching

Presence and condition of appropriate devices

It should be confirmed that devices, such as single-pole or multi-pole switches, installed to control or vary the supply of electrical energy for normal operating purposes are appropriate in quantity, type, function and location. Such devices should be selected in accordance with Table 537.4 of *BS 7671* and be suitable for the most onerous duty they are intended to perform (Regulation 537.3.1.2 refers).

Where a supply changeover of alternative sources of supply is performed by a functional switching device, the device must switch off all live conductors according to Regulation 463.1.4.

Correct operation verified (functional check)

The correct operation of functional devices should be verified, as required by Regulation 643.10. As stated in Regulation 537.3.1.3, some devices used for functional purposes, such as semiconductor devices (electronic controllers/dimmers and the like) may interrupt the current without opening the corresponding poles.

6.10 Current-using equipment (*permanently connected*)

Permanently connected current-using equipment must be inspected to verify that it does not present a risk of fire or shock. The following items are included on the inspection schedule.

10. Current-using equipment (permanently connected)		
10.1	Suitability of equipment in terms of IP and fire ratings:	(.........)
10.2	Enclosure not damaged / deteriorated during installation so as to impair safety:	(.........)
10.3	Suitability for the environment and external influences:	(.........)
10.4	Security of fixing:	(.........)
10.5	Cable entry holes in ceilings above luminaires, sized or sealed so as to restrict the spread of fire:	(.........)

Fig 6.10.1 Current-using equipment

Suitability of equipment in terms of IP and fire ratings

Inspection should confirm that the installed equipment provides adequate protection against contact with live parts (Regulation 416.2 refers), and does not constitute a fire hazard.

The fixed equipment should comply with the relevant requirements of Section 421 of *BS 7671* for protection against fire. The normal operating temperature of fixed electrical equipment should not cause a fire, and the heat generated shall not be dangerous or harmful to adjacent materials. Where there is a risk of temperatures generated by fixed equipment causing a fire hazard one or more of the installation methods detailed in Regulation 421.1.2 should be employed as appropriate.

Note: To minimise the risk of fire, consumer units and similar switchgear installed in domestic (household) premises should satisfy the requirements of Regulation 421.1.201.

Enclosure not damaged/deteriorated during installation so as to impair safety

It should be confirmed that the installed equipment has not suffered damage during the installation process, which may increase the risk of contact with live parts (Regulation 134.1.2 refers).

Suitability for the environment and external influences

Inspection should be carried out during construction and on completion to confirm that the installed equipment is suitably selected and erected to operate safely, given the external influences reasonably likely to occur (Regulation 512.2.1 refers). External influences are listed in Section 522 of *BS 7671*.

If equipment does not meet these requirements by its own construction it must be provided with additional protection, such as by locating in an enclosure, cabinet or the like (Regulation 512.2.2 refers). However, where this is the case, it must be confirmed that the operation of the equipment is not adversely affected by the additional protection.

Security of fixing

Equipment should be securely fixed using appropriate methods and taking into account the equipment manufacturers instructions (Regulation 134.1.1 refers). Regulation Group 530.4 contains requirements for the fixing of equipment. Whilst it is permitted for certain equipment to be installed in or on a cable trunking, skirting trunking or in mouldings, the equipment should not be fixed to covers that can be removed inadvertently (Regulation 530.4.3 refers).

Regulation 559.5.2 requires that adequate means of fixing is provided for luminaires. Where the means of fixing is intended to support the luminaire it should be capable of carrying 5 kg.

Cable entry holes in ceilings above luminaires, sized or sealed so as to restrict the spread of fire

Holes made in the fabric of the building should be sized to restrict the spread of fire or where necessary, the holes should be sealed (refer to 'Provision of fire barriers, sealing arrangements so as to minimise the spread of fire', under Section 8 of the schedule).

Recessed luminaires (downlighters)

Where an installation contains recessed luminaires, the following inspections are required to verify that the luminaires will not present a risk of fire:

10.6	Recessed luminaires (downlighters):	
	a) Correct type of lamps fitted	(.........)
	b) Installed to minimise build-up of heat	(.........)
10.7	Provision of undervoltage protection, where specified:	(.........)
10.8	Provision of overload protection, where specified:	(.........)
10.9	Adequacy of working space / accessibility to equipment:	(.........)

Fig 6.10.2 Current-using equipment

- **Correct type of lamps fitted:** Inspection of lamps is required to confirm that they are the correct type and rating for the luminaire. For example, if a dichroic lamp is inserted into a recessed downlighter the heat generated from the lamp operation will be projected backwards into the ceiling void rather than away from it, creating a risk of fire. Because of such risks, manufacturers place warnings on fittings that are **not** suitable for use with cool-beam lamps.

- **Installed to minimise the build-up of heat:** Downlighters should be installed so that the build-up of heat is minimised. For example, a luminaire mounted on a flammable surface, such as a timber panelled ceiling, should be a type suitable for mounting on such a surface otherwise the heat produced from the luminaire in normal operation may, over time, cause ignition of the surface. Likewise, a luminaire recessed in a ceiling may be covered by thermal insulation, and as such unable to dissipate heat effectively and subsequently overheat. Therefore a method of protecting the fitting from contact with combustible materials, such an insulation displacement box, may be necessary.

The installation of recessed luminaires should not adversely affect the fire resistance required for any part of an installation. For example, where an office ceiling (plasterboard or similar) is required, to provide a minimum fire resistance of **30 minutes** in accordance with building regulations, any luminaire (or other item of electrical equipment such as a speaker) recessed into the ceiling must meet that requirement. Inspection should confirm that the the appropriate 'fire rating' is maintained. (For these reasons, 'fire-rated' luminaires are recommended.)

The selection and erection of a luminaire should take into account the thermal effects of radiant and convected energy on the surroundings. As a result, Regulation 559.4.1 requires the following points to be taken into account:

Fig 6.10.3 Type and selection of lights and lamps suitable for location and position

- the maximum power dissipation of the lamps
- the fire-resistance of adjacent material
- the minimum distance to combustible materials, including material in the path of a spotlight
- the relevant markings on the luminaire.

Manufacturer's provide marking on luminaires and controlgear to indicate that the product is not suitable for particular mounting or installation methods. Table 55.3 of *BS 7671* provides an explanation of *BS EN 60598-1* symbols applicable to luminaires and controlgear.

Note: A luminaire designed to be ground recessed should be selected and erected in accordance with the manufacturer's instructions taking into account the guidance given in Table A.1 of *BS EN60598-2-13* (Regulation 559.10 refers).

BS EN 60598-1:2008	
	Recessed luminaire not suitable for direct mounting on normally flammable surfaces
	Surface mounted luminaire not suitable for direct mounting on normally flammable surfaces
	Luminaire not suitable for covering with thermally insulating material

Fig 6.10.4 Symbols used for marking luminaires and controlgear

Guidance on the impact of electrical installations on the fire performance of domestic premises used as single family houses can be found in Best Practice Guide 5, issued by Electrical Safety First. The Guide can be downloaded from: **www.electricalsafetyfirst.org.uk**

Provision of undervoltage protection

Undervoltage protective devices are provided to disconnect equipment in the event of a significant reduction in the supply voltage. Inspection should confirm undervoltage protection is provided according to the design requirements of the installation and Section 445 of *BS 7671*.

Any necessary arrangements to prevent automatic reclosure of undervoltage devices should also be verified. Such arrangements are required where the reclosure of an undervoltage device is likely to cause danger, for example, where the unexpected restarting of an electric motor could cause danger (Regulation 552.1.3 refers).

Provision of overload protection, where specified

The type and rating of overload protection provided for current-using equipment should be checked against the design and the requirements of Section 433.

For example, the inspector should confirm that the current-settings of motor overloads have been adjusted and installed in accordance with the requirements of *BS 7671* and the electrical designers intentions (Regulation 643.10 refers).

Adequacy of working space / accessibility to equipment

It must be confirmed that adequate and safe means of access and working space have been afforded to every piece of electrical equipment and switchgear to facilitate operation, inspection, testing, fault detection, maintenance and repair (Regulation 513.1 refers).

6.11 Special installations or locations

Any 'Special installation or location' forming part of the inspection should be recorded in this section of the schedule and full details should be provided on a separate numbered page and appended to the certificate.

It should be noted that the particular requirements contained in Part 7 of *BS 7671* supplement or modify the general requirements contained in the other parts of *BS 7671*. Therefore, in addition to verifying compliance with the general requirements of *BS 7671*, the particular requirements of the relevant Section of Part 7 must also be satisfied (Regulation 642.3 refers). Compliance is confirmed by entering a tick in the bracket on the schedule.

11. Special installations or locations

List below any special installations or locations which are part of the installation to be verified, and confirm that the additional requirements given in the respective section of Part 7 are fulfilled:

.... (.........)

.... (.........)

.... (.........)

.... (.........)

Details must be appended on a separate numbered page (see PART 10 below)

Fig 6.11.1 Special installations or locations

A location containing a bath or shower is normally applicable for every domestic installation. For these reasons, the 'Schedule of Items Inspected' provided as part of an NICEIC or ELECSA Domestic Electrical Installation Certificate includes items applicable for a location containing a bath or shower.

It should be noted that for a location containing a bath or a shower, the special location applies to the zones only. For areas outside the defined zones shown in Fig 6.11.2, the normal requirements of *BS 7671* should be applied.

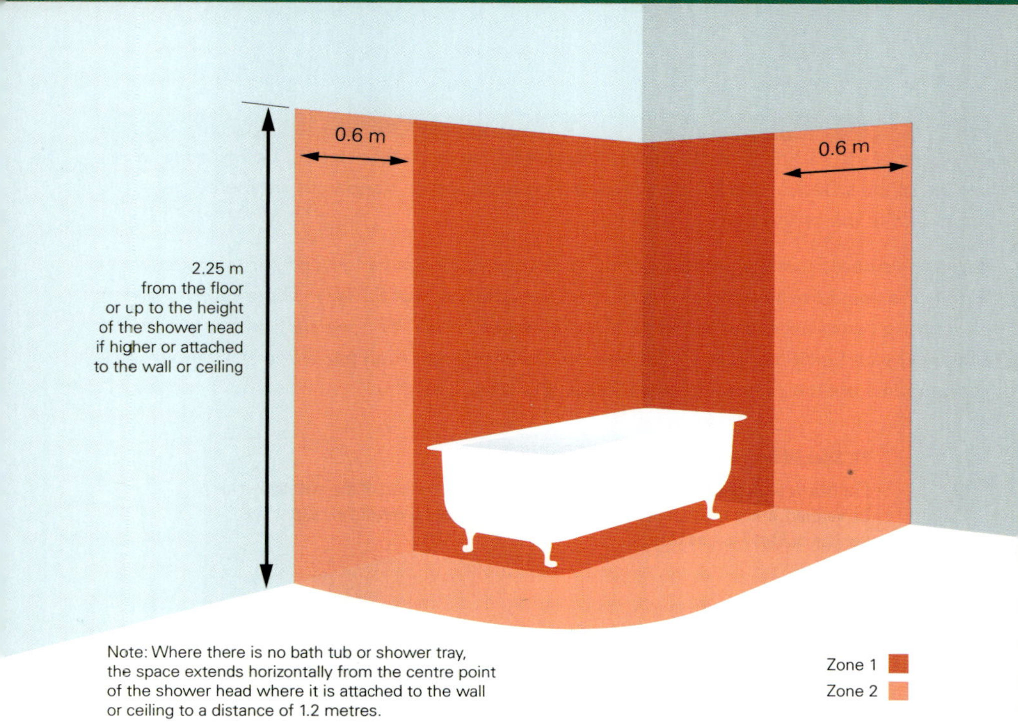

Fig 6.11.2 Location containing a bath or shower

Note: Zone 0 is the interior of the bathtub or shower basin (Regulation 701.32.2 refers).

Schedule of items inspected by

The person who carried out the inspection must sign the completed schedule. In doing so the person is confirming that the record of items inspected is accurate and has been undertaken by a skilled person competent to verify the requirements of *BS 7671* have been met (Regulation 644.5 refers).

CHAPTER 7
THE INSPECTION SCHEDULE FOR CONDITION REPORTING

d:

so as to im

and external inf

8.5 Security of fixing: (..........)

8.6 Cable entry holes in ceiling above luminaires, so as to restrict the spread of fire: (..........) (..........)

List number and location of luminaires inspected on a separate page: (..........) (..........)

kings: (..........)

8.7 Recessed luminaires (e.g. downlighters)

cannot

a) Correct type of lamps fitted: (..........)

b) Installed to minimise build-up of heat:

c) No signs of overheating to surrounding bu

d) No signs of overheating to conductors / te (..........)

9. **List all special installations or locations covered** (..........)

.... (..........)

.... (..........)

king(s): (..........)

....

....

Indicate if the relevant requirements of Part 7 are satisfie (..........)

nt occur: (..........)

of inspection on a separate numbered page. (..........)

SCHEDULE OF ITEMS INSPECTED BY

Name (capitals): .... (..........)

Signature: .... (..........)

7 (i) General

Electrical installations deteriorate with age, as well as with wear and tear. Every Installation, therefore, needs to be inspected at appropriate intervals during its service life, to check that its condition is such that it is safe to remain in service, and is likely to remain safe at least until the next inspection is due. Periodic inspection also enables the need for any remedial work to be identified and drawn to the attention of those responsible for the safety of the installation.

As required by Regulation 653.2, schedules of inspections must be included in the Report issued to the person ordering the inspection. This Chapter provides guidance on completing the 'Schedule of Items Inspected' which is included as a part of the NICEIC Electrical Installation Condition Report.

NICEIC APPROVED CONTRACTOR

This report is not valid if the serial number has been defaced or altered | IPN18

ELECTRICAL INSTALLATION CONDITION REPORT

Issued in accordance with BS 7671: 2018 – Requirements for Electrical Installations

Original (to the person ordering the work)

PART 10 : SCHEDULE OF ITEMS INSPECTED

1. External condition of electrical intake equipment (visual inspection only)
(If inadequacies are identified with the intake equipment, it is recommended the person ordering the report informs the appropriate authority.)

1.1 Service cable: (.........) 1.2 Service head: (.........)
1.3 Earthing arrangement: (.........) 1.4 Meter tails: (.........)
1.5 Metering equipment: (.........) 1.6 Isolator (where present): (.........)

2. Presence of adequate arrangements for parallel or switched alternative sources
2.1 Adequate arrangements where a generating set operates as a switched alternative to the public supply: (.........)
2.2 Adequate arrangements where generating set operates in parallel with the public supply: (.........)
2.3 Presence of alternative / additional supply arrangement warning notice(s) at or near equipment, where required: (.........)

3. Automatic disconnection of supply
3.1 Main earthing and bonding arrangements
a) Presence and condition of distributor's earthing arrangement: (.........)
b) Presence and condition of earth electrode arrangement, if present: (.........)
c) Adequacy of earthing conductor size: (.........)
d) Adequacy of earthing conductor connections: (.........)
e) Accessibility of earthing conductor connections: (.........)
f) Adequacy of main protective bonding conductor size(s): (.........)
g) Adequacy of main protective bonding conductor connections: (.........)
h) Accessibility of main protective bonding connections: (.........)
i) Accessibility and condition of other protective bonding connections: (.........)
j) Provision of earthing / bonding labels at all appropriate locations: (.........)

3.2 FELV
a) Source providing at least simple separation: (.........)
b) Plugs, socket-outlets and the like not interchangeable with those of other systems within the premises: (.........)

4. Other methods of protection (.........)
Details should be provided on separate sheets: Page No. (.........)

5. Distribution equipment
5.1 Adequacy of working space / accessibility of equipment: (.........)
5.2 Security of fixing: (.........)
5.3 Condition of insulation of live parts: (.........)
5.4 Adequacy / security of barriers: (.........)
5.5 Condition of enclosure(s) in terms of IP rating: (.........)
5.6 Condition of enclosure(s) in terms of fire rating: (.........)
5.7 Enclosure not damaged / deteriorated so as to impair safety: (.........)
5.8 Presence and effectiveness of obstacles: (.........)
5.9 Presence of main switch(es), linked where required: (.........)
5.10 Operation of main switch(es) *(functional check)*: (.........)
5.11 Correct identification of circuit protective devices: (.........)
5.12 Adequacy of protective devices for prospective fault current: (.........)
5.13 RCD(s) provided for fault protection – includes RCBOs: (.........)
5.14 RCD(s) provided for additional protection – includes RCBOs: (.........)
5.15 RCD(s) provided for protection against fire – includes RCBOs: (.........)
5.16 Manual operation of circuit-breakers and RCDs to prove disconnection: (.........)
5.17 Confirmation that integral test button/switch causes RCD(s) to trip when operated (functional check): (.........)
5.18 Presence of RCD six-monthly retest notice at or near equipment, where required: (.........)
5.19 Presence of diagrams, charts or schedules at or near equipment, where required: (.........)
5.20 Presence of non-standard (mixed) cable colour warning notices at or near equipment, where required: (.........)
5.21 Presence of next inspection recommendation label: (.........)
5.22 All other required labelling provided: (.........)
5.23 Compatibility of protective device(s), base(s) and other components: (.........)
5.24 Single-pole switching or protective devices in line conductors only: (.........)
5.25 Protection against mechanical damage where cables enter equipment: (.........)
5.26 Protection against electromagnetic effects where cables enter ferrromagnetic enclosures: (.........)

6. Distribution / final circuits
6.1 Identification of conductors: (.........)
6.2 Cables correctly supported throughout their length: (.........)
6.3 Condition of insulation of live parts: (.........)
6.4 Non-sheathed cables protected by enclosures in conduit, ducting or trunking: (.........)
6.5 Suitability of containment systems for continued use *(including flexible conduit)*: (.........)
6.6 Cables correctly terminated in enclosures *(indicate extent of sampling in PART 7 of report)*: (.........)
6.7 Indication of SPD(s) continued functionality confirmed: (.........)
6.8 Adequacy of AFDD(s), where specified: (.........)
6.9 Confirmation that conductor connections, including connections to busbars are correctly located in terminals and are tight and secure: (.........)
6.10 Examination of cables for signs of unacceptable thermal and mechanical damage / deterioration: (.........)
6.11 Adequacy of cables for current-carrying capacity with regard to the type and nature of installation: (.........)
6.12 Adequacy of protective devices; type and rated current for fault protection: (.........)
6.13 Presence and adequacy of circuit protective conductors: (.........)
6.14 Co-ordination between conductors and overload protective devices: (.........)
6.15 Cable installation methods / practices appropriate to the type and nature of installation and external influences: (.........)
6.16 Cables where exposed to direct sunlight, of a suitable type or adequately protected against solar radiation: (.........)
6.17 Cables adequately protected against damage and abrasion: (.........)

All fields must be completed. *Enter either, as appropriate: '✓' if Acceptable condition; '**N/A**' if Not applicable; '**LIM**' if a Limitation exists; or Code appropriately – CODE '**C1**', '**C2**', '**C3**' or '**FI**' (codes to be recorded in PART 6, with additional comments (where appropriate) on attached numbered sheets)*

IPN18/7

This report is based on the model forms shown in Appendix 6 of *BS 7671*
Published by Certsure LLP Certsure LLP operates the NICEIC & ELECSA brands © Copyright Certsure LLP (July 2018)
Warwick House, Houghton Hall Park, Houghton Regis, Dunstable, LU5 5ZX

Page 4 of

Fig 7 (i) Schedule of Items Inspected

7.0 Completion of the 'Schedule of Items Inspected'

It should be appreciated that the 'Schedule of Items Inspected' which is based on the model contained in Appendix 6 of *BS 7671*, is not intended to be an exhaustive list. Depending on the particular installation being inspected some of the items listed may not be applicable or alternatively in some circumstances additional items may need to be included (on a separate page). However, in all cases an entry should be made for each item listed on the Schedule using one of the following options:

(√) the result was satisfactory.

C1 an unsatisfactory condition was observed, such that danger was present.

C2 an unsatisfactory condition was observed which was potentially dangerous.

C3 means improvement recommended and should be used to indicate that, whilst an observed deficiency was not considered to be a source of immediate or potential danger, improvement would contribute to improving the safety of the electrical installation.

F I indicates that further investigation (without delay) is required to determine whether danger or potential danger exists.

Lim indicates that, exceptionally, a limitation agreed with the person ordering the work prevented the inspection being carried out.

N/A means that an item on the Inspection Schedule was not applicable to the particular installation.

Note: If further investigation is required, the condition of the installation must be reported to be unsatisfactory.

The remainder of this Chapter looks at the items contained in the 'Schedule of Items Inspected' which forms part of the NICEIC Electrical Installation Condition Report. Many of the items included are also contained in the Schedule which forms part of the NICEIC Electrical Installation Certificate so where appropriate the reader is referred to the guidance provided in Chapter 6.

7.1 External condition of intake equipment (visual inspection only)

Where accessible the supply intake equipment including the meter tails should be inspected for signs of mechanical damage, evidence of overheating and other forms of deterioration. Where signs of significant damage are observed these should be noted in Part 6 of the Report 'Observations and recommendations for actions to be taken'. The person who ordered the inspection should be advised to inform the distributor and / or meter operator, as appropriate, of any inadequacies in the equipment forming a part of their installation that have been observed during the inspection.

The suppliers earthing arrangement should be inspected to confirm it is suitable for the particular supply system type, TN-S, TN-C-S or TT being the most common systems used. (Chapter 6, 'Automatic disconnection of supply 'provides details on these types of supply systems.)

The type and rating of the service fuse(s) should be ascertained by reference to the original design or specification. The fuse rating will be important for determining the adequacy of the meter tails and should be entered in the appropriate field in Part 8 of the Report 'Supply characteristics and earthing arrangements'.

Under no circumstances should the inspector attempt to access or withdraw the cut-out fuse(s) as a part of this inspection, unless prior authorisation has been obtained.

1. External condition of electrical intake equipment (visual inspection only)
(If inadequacies are identified with the intake equipment, it is recommended the person ordering the report informs the appropriate authority.)

1.1	Service cable:	(.........)	1.2	Service head:	(.........)
1.3	Earthing arrangement:	(.........)	1.4	Meter tails:	(.........)
1.5	Metering equipment:	(.........)	1.6	Isolator (where present):	(.........)

Fig 7.1 Intake equipment

7.2 Presence of adequate arrangements for parallel or switched alternative sources

Where there are parallel or switched sources of supply to, or within, the installation, the inspector is required to confirm that the connection and operation of the generating set complies with the requirements detailed in Regulation 551.6 and Regulation 551.7 as appropriate.

2. Presence of adequate arrangements for parallel or switched alternative sources

2.1	Adequate arrangements where a generating set operates as a switched alternative to the public supply:	(.........)
2.2	Adequate arrangements where generating set operates in parallel with the public supply:	(.........)
2.3	Presence of alternative / additional supply arrangement warning notice(s) at or near equipment, where required:	(.........)

Fig 7.2 Presence of adequate arrangementsfor parallel or switched alternative sources

Where automatic disconnection cannot be achieved in the time required by Chapter 41 of *BS 7671* for equipment, such as an electronic power convertor, having a nominal voltage greater than 50 V AC or 120 V DC, it must be confirmed that under earth fault conditions the output voltage of the source will reduce to less than 50 V AC or 120 V DC within the appropriate disconnection time of Chapter 41 (Regulation 419.2 refers). Where this cannot be achieved supplementary bonding should be provided in accordance with Regulation 419.3. (Details on the requirements for parallel or switched alternative sources of supply are covered in Chapter 6.)

7.3 Automatic disconnection of supply

The majority of electrical installations within the UK rely primarily on automatic disconnection of supply (ADS) for protection against electric shock. In view of this, Section 3.1 requires confirmation of the presence, condition and adequacy of the main earthing and protective bonding arrangements for the installation. (Earthing and bonding is covered in Chapter 6.)

In addition, it should be confirmed that the labelling required by Regulation 514.13.1 has been provided at the relevant points within the installation.

3. Automatic disconnection of supply

3.1 Main earthing and bonding arrangements

- a) Presence and condition of distributor's earthing arrangement: (.........)
- b) Presence and condition of earth electrode arrangement, if present: (.........)
- c) Adequacy of earthing conductor size: (.........)
- d) Adequacy of earthing conductor connections: (.........)
- e) Accessibility of earthing conductor connections: (.........)
- f) Adequacy of main protective bonding conductor size(s): (.........)
- g) Adequacy of main protective bonding conductor connections: (.........)
- h) Accessibility of main protective bonding connections: (.........)
- i) Accessibility and condition of other protective bonding connections: (.........)
- j) Provision of earthing / bonding labels at all appropriate locations: (.........)

3.2 FELV

- a) Source providing at least simple separation: (.........)
- b) Plugs, socket-outlets and the like not interchangeable with those of other systems within the premises: (.........)

Fig 7.3 Automatic disconnection of supply

Section 3.2 is only applicable where functional extra-low voltage (FELV) has been employed within an installation as part of the protective measure of automatic disconnection of supply (ADS). For details on FELV refer to Chapter 6.

7.4 Other methods of protection

If a method of protection other than ADS is employed it should be indicated by inserting a 'tick' in the field. Details of the protective measure used along with confirmation of any inspection aspects relating to its use should be provided on an additional numbered page and attached to the report. (Protective measures are covered in Chapter 6 of this guide.)

4. Other methods of protection (.........)

Details should be provided on separate sheets: Page No. (.........)

Fig 7.4 Other methods of protection

Where ADS is the sole method of protection for the installation being inspected, N/A should be entered.

7.5 Distribution equipment

The items listed in this section of the schedule are specific to distribution equipment, and the inspections undertaken should be appropriate for the size and type of installation. The items are generally self-explanatory and closely follow those detailed in Chapter 6, so a general summary of the items included is provided as follows:

5. Distribution equipment		
5.1	Adequacy of working space / accessibility of equipment:	(.........)
5.2	Security of fixing:	(.........)
5.3	Condition of insulation of live parts:	(.........)
5.4	Adequacy / security of barriers:	(.........)
5.5	Condition of enclosure(s) in terms of IP rating:	(.........)
5.6	Condition of enclosure(s) in terms of fire rating:	(.........)
5.7	Enclosure not damaged / deteriorated so as to impair safety:	(.........)
5.8	Presence and effectiveness of obstacles:	(.........)

Fig 7.5.1 Distribution equipment

These items require the distribution equipment to be inspected to confirm that it remains securely fixed and has not suffered damage or deterioration which has compromised the level of protection it is required to provide, including fire and IP ratings. For operational and maintenance purposes such equipment must be accessible (Regulation 513.1 refers).

5.9	Presence of main switch(es), linked where required:	(.........)
5.10	Operation of main switch(es) *(functional check):*	(.........)
5.11	Correct identification of circuit protective devices:	(.........)
5.12	Adequacy of protective devices for prospective fault current:	(.........)
5.13	RCD(s) provided for fault protection – includes RCBOs:	(.........)
5.14	RCD(s) provided for additional protection – includes RCBOs:	(.........)
5.15	RCD(s) provided for protection against fire – includes RCBOs:	(.........)
5.16	Manual operation of circuit-breakers and RCDs to prove disconnection:	(.........)
5.17	Confirmation that integral test button/switch causes RCD(s) to trip when operated (functional check)	(.........)

Fig 7.5.2 Distribution equipment

These items relate to the devices installed within the equipment such as main switches, protective devices and RCBOs. The inspector is required to confirm that devices installed to provide fault protection, additional and fire protection remain suitable for their intended function. This includes performing functional tests to prove the devices will disconnect.

The following items are straightforward. The first part of these items relates to the requirements of Section 514 of *BS 7671* for the provision of identification and notices, whilst among the remaining items an examination of protective devices for signs of thermal or mechanical damage is required by item 5.23.

5.18	Presence of RCD six-monthly retest notice at or near equipment, where required:	(.........)
5.19	Presence of diagrams, charts or schedules at or near equipment, where required:	(.........)
5.20	Presence of non-standard (mixed) cable colour warning notices at or near equipment, where required:	(.........)
5.21	Presence of next inspection recommendation label:	(.........)
5.22	All other required labelling provided:	(.........)
5.23	Compatibility of protective device(s), base(s) and other components:	(.........)
5.24	Single-pole switching or protective devices in line conductors only:	(.........)
5.25	Protection against mechanical damage where cables enter equipment:	(.........)
5.26	Protection against electromagnetic effects where cables enter ferrromagnetic enclosures:	(.........)

Fig 7.5.3 Distribution equipment

7.6 Distribution / final circuits

This section of the schedule lists 27 items, most of which are self-explanatory, relating to distribution and final circuits.

6. Distribution / final circuits		
6.1	Identification of conductors:	(.........)
6.2	Cables correctly supported throughout their length:	(.........)
6.3	Condition of insulation of live parts:	(.........)
6.4	Non-sheathed cables protected by enclosures in conduit, ducting or trunking:	(.........)
6.5	Suitability of containment systems for continued use *(including flexible conduit):*	(.........)
6.6	Cables correctly terminated in enclosures *(indicate extent of sampling in PART 7 of report):*	(.........)
6.7	Indication of SPD(s) continued functionality confirmed:	(.........)
6.8	Adequacy of AFDD(s), where specified:	(.........)
6.9	Confirmation that conductor connections, including connections to busbars are correctly located in terminals and are tight and secure:	(.........)

Fig 7.6.1 Distribution / final circuits

As required by item 6.6, sufficient inspection should be undertaken to verify that connections to enclosures have adequate mechanical protection and protection against relevant external influences (Regulation 526.7 refers). For such purposes, the level of sampling should be appropriate for the size and type of installation and should have been agreed with the person ordering the Report prior to the inspection. Where defects are identified the degree of sampling should be increased appropriately; where necessary all equipment enclosures may need to be checked. Likewise, an appropriate level of sampling is required for the tightness and security of conductor connections to equipment and busbars.

The items shown are self-explanatory and relate to the condition and adequacy of cables and their associated protective devices.

6.10	Examination of cables for signs of unacceptable thermal and mechanical damage / deterioration:	(.........)
6.11	Adequacy of cables for current-carrying capacity with regard to the type and nature of installation:	(.........)
6.12	Adequacy of protective devices; type and rated current for fault protection:	(.........)
6.13	Presence and adequacy of circuit protective conductors:	(.........)
6.14	Co-ordination between conductors and overload protective devices:	(.........)
6.15	Cable installation methods / practices appropriate to the type and nature of installation and external influences:	(.........)
6.16	Cables where exposed to direct sunlight, of a suitable type or adequately protected against solar radiation:	(.........)
6.17	Cables adequately protected against damage and abrasion:	(.........)

Fig 7.6.2 Distribution / final circuits

As stated on the schedule, older installations designed prior to *BS 7671: 2008* may not have been provided with RCDs for additional protection. However, such installations according to their design may still be safe for continued use. Where this is the case, a code C3 should be assigned to identify to the person ordering the work that the installation would benefit from RCD protection.

6.18	Provision of additional protection by an RCD not exceeding 30 mA	
a)	For all socket-outlets with a rated current not exceeding 32 A, unless exempt:	(.........)
b)	Supplies for mobile equipment with a rated current not exceeding 32 A for use outdoors:	(.........)
c)	For cables concealed in walls / partitions at a depth of less than 50 mm:	(.........)
d)	For cables concealed in walls / partitions containing metal parts regardless of depth:	(.........)
e)	Circuits supplying luminaires within domestic (household) premises:	(.........)

Fig 7.6.3 Provision of additional protection by an RCD not exceeding 30 mA

Of the remaining items, 6.22 contains four inspections specific to the termination of cables. The extent of sampling applied for this item should appropriate and indicated in PART 7 of the Report.

6.19	Provision of fire barriers, sealing arrangements and protection against thermal effects:	(.........)
6.20	Band II cables segregated / separated from Band I cables:	(.........)
6.21	Cables segregated / separated from non-electrical services:	(.........)
6.22	Termination of cables at enclosures *(indicate extent of sampling in PART 7 of report)*	
	a) Connections under no undue strain:	(.........)
	b) No basic insulation of a conductor, visible outside an enclosure:	(.........)
	c) Connections of live conductors adequately enclosed:	(.........)
	d) Adequacy of connection at point of entry to enclosure:	(.........)
6.23	Temperature rating of cable insulation addequate:	(.........)
6.24	Condition of accessories including socket-outlets, switches and joint boxes satisfactory:	(.........)
6.25	Suitability of accessories for external influences:	(.........)
6.26	Single-pole switching or protective devices in line conductors only:	(.........)
6.27	Adequacy of connections, including cpcs, within accessories and to fixed and stationary equipment:	(.........)

Fig 7.6.4 Distribution / final circuits

7.7 Isolation and switching

This section has four parts listing a number of inspection items relating to devices providing:

- isolation
- switching off for mechanical maintenance
- emergency switching / stopping
- functional switching.

7. Isolation and switching

7.1 Isolators
- a) Presence and condition of appropriate devices:
- b) Acceptable location (local / remote):
- c) Capable of being secured in the OFF position:
- d) Correct operation verified:
- e) Clearly identified by position and / or durable markings:
- f) Warning label posted in situations where live parts cannot be isolated by the operation of a single device:

7.2 Switching off for mechanical maintenance
- a) Presence and condition of appropriate devices:
- b) Acceptable location:
- c) Capable of being secured in the OFF position:
- d) Correct operation verified:
- e) Clearly identified by position and / or durable marking(s):

7.3 Emergency switching off / stopping
- a) Presence and condition of appropriate devices:
- b) Readily accessible for operation where danger might occur:
- c) Correct operation verified:

7.4 Functional switching
- a) Presence and condition of appropriate devices:
- b) Correct operation (functionality) verified:

Fig 7.7 Isolation and switching

It should be noted that a single device or collection of devices may provide more than one of the functions listed above. Where this is the case it will be necessary that the device meets all the relevant requirements of *BS 7671* for each of the functions it is intended to provide. In such situations, prior knowledge of the installation and / or availability of relevant information relating to the design of the installation will be highly beneficial. Guidance on isolation and switching is given in Chapter 6.

7.8 Current-using equipment (permanently connected)

This section of the schedule calls for the inspection of permanently connected current-using equipment (including electrical equipment present in accessible roof spaces) to verify that it does not present a hazard in terms of shock or fire.

8. Current-using equipment *(permanently connected)*		
8.1	Condition of equipment in terms of IP rating:	(.........)
8.2	Equipment does not constitute a fire hazard:	(.........)
8.3	Enclosure not damaged / deteriorated so as to impair safety:	(.........)
8.4	Suitability for the environment and external influences:	(.........)
8.5	Security of fixing:	(.........)
8.6	Cable entry holes in ceiling above luminaires, sized or sealed so as to restrict the spread of fire:	(.........)

Fig 7.8.1 Current-using equipment (*permanently connected*)

Equipment enclosures installed as part of the fixed installation should be inspected to confirm their continued adequacy in terms of IP rating and to identify any damage / deterioration that has occurred whilst in service. In particular, where equipment such as motors, pumps and fans are not regularly maintained it can, over time, result in deterioration and thermal damage.

List number and location of luminaires inspected on a separate page:		Page No. (.........)
8.7	Recessed luminaires (e.g. downlighters)	
	a) Correct type of lamps fitted:	(.........)
	b) Installed to minimise build-up of heat:	(.........)
	c) No signs of overheating to surrounding building fabric:	(.........)
	d) No signs of overheating to conductors / terminations:	(.........)

Fig 7.8.2 Current-using equipment

Where the installation contains recessed luminaires, a sample inspection should be made to verify that the luminaire is suitable for the location and lamps currently fitted are of the correct type and rating. In agreement with the person ordering the inspection, some recessed luminaires should be withdrawn to inspect the fitting and the cavity for signs of thermal damage, such as scorching of joists and similar structures. Details of the inspection sample should be provided on an additional numbered page and appended to the Report.

Whilst no additional measures may be necessary in the case of 'fire-rated' luminaires, where open backed type luminaires have been installed, an insulation displacement box or similar may be necessary.

7. 9 Special installations or locations covered by this Report

All special installations or locations covered by Part 7 of *BS 7671*, including locations containing a bath or shower, should be recorded in this section. By inserting a 'tick' in the appropriate field the inspector is confirming that the particular requirements of the relevant section of PART 7 have been satisfied in addition to the general requirements of *BS 7671*.

The details of the particular inspections undertaken should be recorded on an additional numbered page and appended to the Report.

9. List all special installations or locations covered by this report:	
....	(.........)
....	(.........)
....	(.........)
....	(.........)
Indicate if the relevant requirements of Part 7 are satisfied and append results of inspection on a separate numbered page.	

Fig 7.9.1 Special installations or locations

The person carrying out the inspection should sign to confirm that the schedule has been completed in accordance with *BS 7671*. The date inserted should be the date (or dates) when the inspection was carried out rather than the date the report was issued.

SCHEDULE OF ITEMS INSPECTED BY	
Name (capitals):..	
Signature:..	Date:............................

Fig 7.9.2 Signature

Any items that have been identified as being dangerous (code C1) or potentially dangerous (code C2), or that would result in a significant enhancement of safety if improved (code C3), or any items that require further investigation without delay to determine whether danger or potential danger exists (code FI), should be carried forward to Part 6 of the Report in the form of observations on the safety of the installation.

CHAPTER 8
TESTING

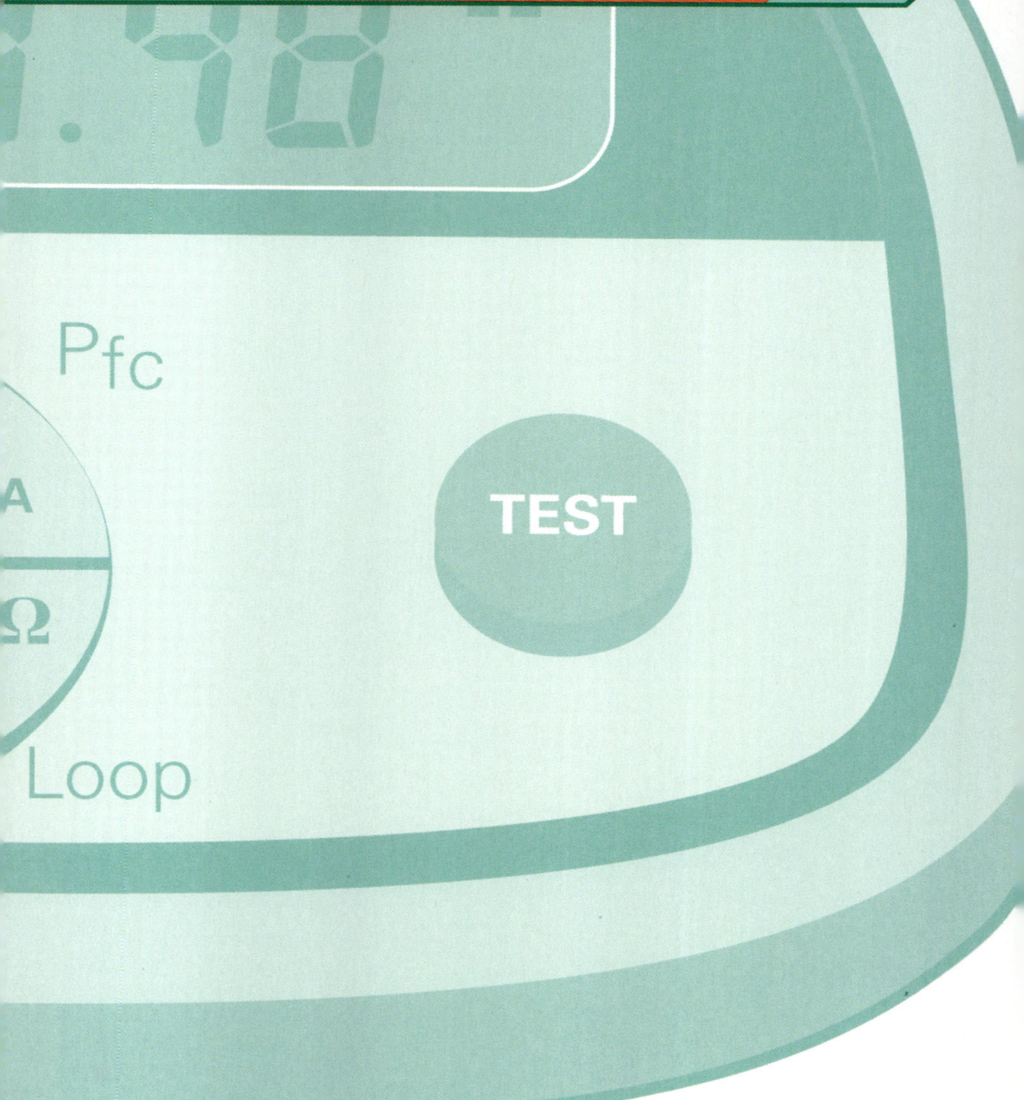

8.1 Safety

Electrical testing inherently involves a degree of risk. Persons carrying out electrical testing have a duty to ensure the safety of themselves and others from any possible danger resulting from the tests or from access to live parts. This requires strict adherence to safe working practices, such as the 'Isolation procedure' (see Appendix 1), and consistently undertaking appropriate checks to confirm safety, such as checking that:

- the test equipment and leads are safe and suitable for the intended purpose,
- the equipment to be tested is safe for the tests, and
- the working environment does not impair safety.

Although for the purposes of 'live' testing, electrical equipment covers, such as distribution board or consumer unit covers, do have to be removed, once the covers are removed the 'live' equipment should never be left unattended, otherwise as shown in Fig 8.1 persons may be exposed to the risk of electric shock.

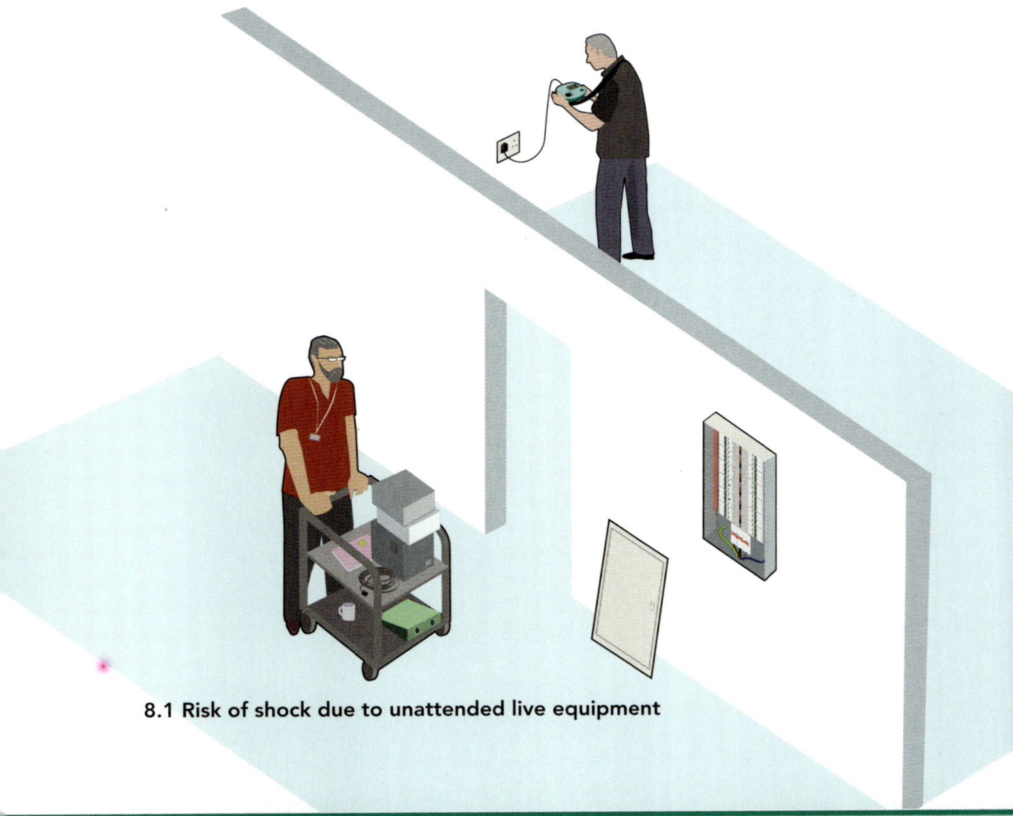

8.1 Risk of shock due to unattended live equipment

Persons carrying out testing, should be aware of the danger present when connecting test instruments to live parts, and take suitable precautions, as required by the *Electricity at Work Regulations 1989* (EWR), so that persons are not placed at risk when carrying out 'live' tests. In particular, where the information required can be obtained by a method other than a test, such as by enquiry, live testing should be avoided. (For guidance on the EWR refer to Chapter 1).

Measuring instruments, monitoring equipment and methods used for electrical testing should conform to the relevant parts of *BS EN 61557**, or provide at least the same level of performance and safety (Regulation 643.1 refers). Safety advice on the use of electrical test equipment is given in HSE Guidance Note GS38: 'Electrical Test Equipment for use on low voltage electrical systems'. For details on GS38 refer to Appendix B of this Guide.

Note: Particular care should be exercised where the installation is connected to additional supply sources, such as those provided by standby generators and uninterruptible power supplies.

8.2 Instrument accuracy

It is essential that contractors have a system in place to confirm and maintain the accuracy and consistency of all the test instruments they use for certification and reporting purposes. Therefore, in addition to formal calibration/re-calibration of test instruments by recognised organisations (preferably those having UKAS accreditation, at the intervals recommended by the instrument manufacturer) contractors should have a system in place for monitoring the ongoing accuracy of test instruments. Undertaking regular checks between formal calibration will help to ensure that any instrument inaccuracies that might arise are identified in a timely fashion. If test instruments are not monitored between formal calibrations, then any instrument inaccuracy that is identified could affect a large number of certificates and reports.

Typically such a system may include:

- maintaining records over time of comparative cross-checks with other test instruments, or
- maintaining records over time of measurements of the characteristics of designated reference circuits or devices. For example, the consistency of continuity and insulation resistance may be checked against a proprietary resistance box 'check box' or a resistance check card (see Fig 8.2 (i)) or a set of suitable resistors.

* *BS EN 61557-16: 2015 Electrical safety in low voltage distribution systems up to 1 000 V a.c. and 1 500 V d.c. — Equipment for testing, measuring or monitoring of protective measures*

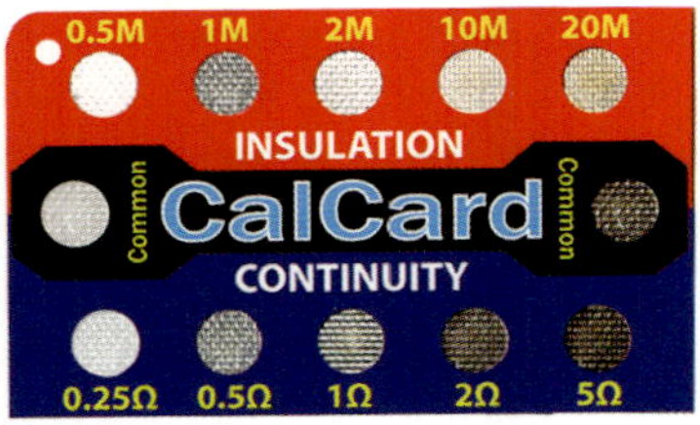

Fig 8.2 (i) NICEIC and ELECSA 'CalCard' – Resistance check card

The accuracy of an earth fault loop impedance and RCD test instrument may be checked by carrying out tests on a designated socket-outlet at regular intervals.

Irrespective of the method used to maintain and monitor accuracy, each test instrument used for testing electrical installations should be clearly and uniquely identified and a log of the test results for each instrument maintained. The interval between the instrument checks should be dependent on how frequently an instrument is used. Where test instruments are used regularly, on a weekly basis, an interval of one or two months is advisable.

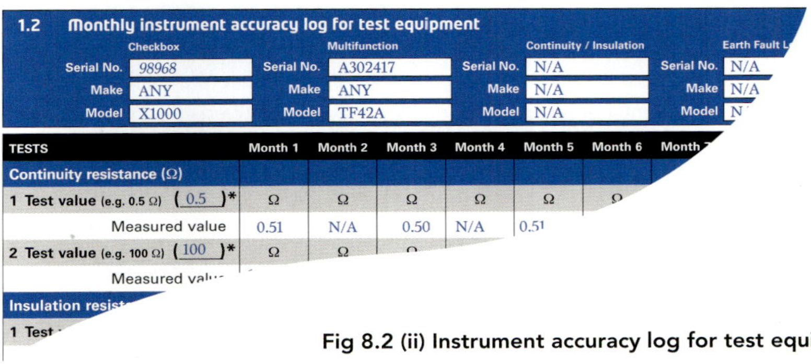

1.2 Monthly instrument accuracy log for test equipment

	Checkbox	Multifunction	Continuity / Insulation	Earth Fault L
Serial No.	98968	A302417	N/A	N/A
Make	ANY	ANY	N/A	N/A
Model	X1000	TF42A	N/A	N

TESTS	Month 1	Month 2	Month 3	Month 4	Month 5	Month 6	Month 7
Continuity resistance (Ω)							
1 Test value (e.g. 0.5 Ω) (0.5)*	Ω	Ω	Ω	Ω	Ω	Ω	
Measured value	0.51	N/A	0.50	N/A	0.51		
2 Test value (e.g. 100 Ω) (100)*	Ω	Ω	Ω				
Measured val...							
Insulation resist...							
1 Test ...							

Fig 8.2 (ii) Instrument accuracy log for test equipment

So that instruments used for testing electrical installations can be easily traced when required the serial numbers of all test instruments are required to be recorded in the appropriate section of NICEIC and ELECSA certificates and reports.

Note: *The Electrical Safety First Best Practice Guide 7* provides guidance for electrical installers on the use of test instruments. The Guide is available to download at:

www.electricalsafetyfirst.org.uk

8.3 Required tests

For the purposes of safety and for practical and operational reasons, the range of tests required and the order which testing is carried out for periodic testing will generally differ from that required for new installation work (including work involving an addition or an alteration). In the case of periodic testing, the installation is in service and generally has been for many years, so the appropriate level of testing should always be agreed with the person ordering the work. The agreed level of testing should be based on experience and knowledge of the type of installation and by consulting available records (guidance on periodic inspection is given in Chapter 4).

For the initial verification of new work the 'dead tests' as detailed in Regulations 643.2 to 643.6 should be carried out in sequence before the installation is energised, and for an installation that contains an earth electrode, a measurement of electrode resistance to Earth (R_A) should also be undertaken before the installation is energised. In circumstances where it is not practicable to perform this 'dead' test, such as where the nature of the ground or space available makes it impracticable to conduct the test, a measurement of external impedance Z_e may be used to verify the resistance of the electrode (Regulation 643.7.2 refers). Guidance on these test methods is given later in this Chapter.

The sequence of tests required for the initial verification of new work, is listed in Fig 8.3. Should a test indicate a failure to comply, that test and any preceding test, the results of which may have been affected by the fault, must be repeated after the fault has been rectified (Regulation 643.1 refers).

If any test carried out during periodic inspection and testing of an installation indicates a failure to comply, then (unless there is evidence of real and immediate danger which requires immediate remedial action as discussed in Chapter 4) the result of that test must be duly recorded on the Electrical Installation Condition Report together with appropriate observations, recommendations and, where appropriate, comments in the summary section of the Report (Regulation 651.4 refers).

The remainder of this Chapter provides guidance on performing the tests listed in Fig 8.3 and verifying the results. It should be noted that methods other than those described in this Chapter are not precluded, and irrespective of the particular test method employed the test instruments used should be operated in accordance with the instrument manufacturer's instructions.

Sequence of testing for initial verification

Dead tests: (undertaken before the supply is connected, or with the supply disconnected as appropriate)

- **Continuity of conductors**

Including

Continuity of protective conductors, main and supplementary equipotential bonding, and the live conductors of ring final circuits.

- **Insulation resistance**

Including: Low voltage circuits and FELV

SELV, PELV or by separation of circuits.

Resistance/impedance of floors and walls

- **Polarity**
- **Earth electrode resistance (where practicable)**

Live tests: (undertaken when the electrical supply is connected)

- **Confirm polarity before further testing**
- **Protection by automatic disconnection of the supply**

Including: Earth fault loop impedance (Z_e and Z_s)

RCD testing

Prospective fault current

- **Verification of phase sequence**
- **Functional testing**
- **Verification of voltage drop (if required)**

Note: For initial verification, the tests performed before connecting the supply (or with the supply disconnected) should be completed in sequence (Regulation 643.1 refers).

Fig 8.3 Sequence of electrical tests for initial verification

8.4 Continuity of conductors

As required by Regulation 643.2.1, resistance measurements should be undertaken to verify the continuity of:

- protective conductors, including protective bonding conductors, and
- live conductors of ring final circuits.

Prior to undertaking continuity 'dead' tests the circuit or supply must be isolated for the purpose of the test, and where access to enclosures containing live parts is required, safe isolation must be carried out prior to gaining access.

For safety reasons, the supply to an installation must be isolated prior to the disconnection of the Earthing conductor or any main protective bonding conductors. Where it is not practicable to isolate the supply, the Earthing and main protective conductors should not be removed.

Test instrument

The results of these continuity tests should be of low ohmic value, and depending on the length and dimensions of the conductors tested are likely to be in the order of milliohms. Therefore, the test instrument to be used is an ohmmeter having a low resistance range, or a multi-function instrument set to the 'continuity' range. In view of this, the resistance of the test leads will be significant, and should not be included in any recorded test result.

Generally, test instruments have provision for correcting for the resistance of the test leads, such as an 'auto-null' facility, but where no such facility is incorporated, it will be necessary to measure and record the resistance of the leads when connected together, and subsequently subtract this value from the test reading.

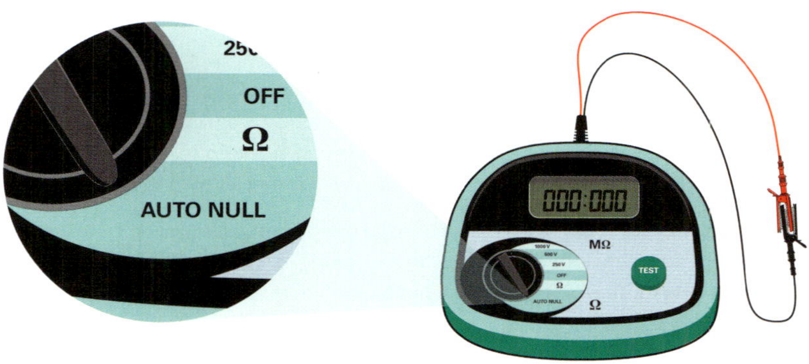

Fig 8.4 Continuity tester incorporating an 'auto null' facility

Parallel paths

It should be appreciated, that on completion of an installation comprised of metal conduit and trunking fixed to metallic structures and / or luminaires fitted to the structural steel, removing parallel paths for the purposes of testing would be impracticable. Consequently, the inspector should be aware of the presence of parallel earth paths and, where possible, testing carried out as part of the initial verification should be undertaken at appropriate times during the installation.

8.4.1 Continuity testing of protective conductors, including protective bonding conductors

It is essential to ensure that all circuit protective conductors and bonding conductors are continuous. Otherwise, an exposed-conductive-part, an extraneous-conductive-part or the earthing terminal of a point or accessory could be left without an effective connection to Earth, giving no fault protection in the event of an earth fault. The test method used to verify continuity will depend on whether the protective conductor is installed separately, such as a main protective bonding conductor, or installed as a part of a circuit.

Verification of continuity test readings may be achieved by comparing measured readings with the tabulated resistance per metre values given in Appendix H of this Guide, based on the particular conductor length and csa.

The R_2 (wander lead) method

This method is used principally for testing main and supplementary protective bonding conductors. Using a long lead (wander lead), the leads of the instrument are attached as shown in Fig 8.4.1 (i), having temporarily disconnected the conductor from the MET to remove parallel paths. In order to confirm a low resistance connection between the *BS 951* bonding clamp and the pipe, the test connection should be made to the pipe and not the conductor.

The wander lead method may also be used to verify the continuity of earth connections made at every position in a circuit, such as earth points at socket-outlets, switches, lighting points, fixed equipment and the like. Using a wander lead, one lead is connected to the main earthing terminal or, as shown in Fig 8.4.1 (ii), to the earth terminal in the consumer unit or distribution board while the other the lead is connected to each earth point within the installation.

It should be noted that the due to the length of the wander lead the resistance of the leads will be significant. It is essential therefore that the resistance of the wander lead is either 'nulled' prior to testing or is subtracted from the value obtained from testing.

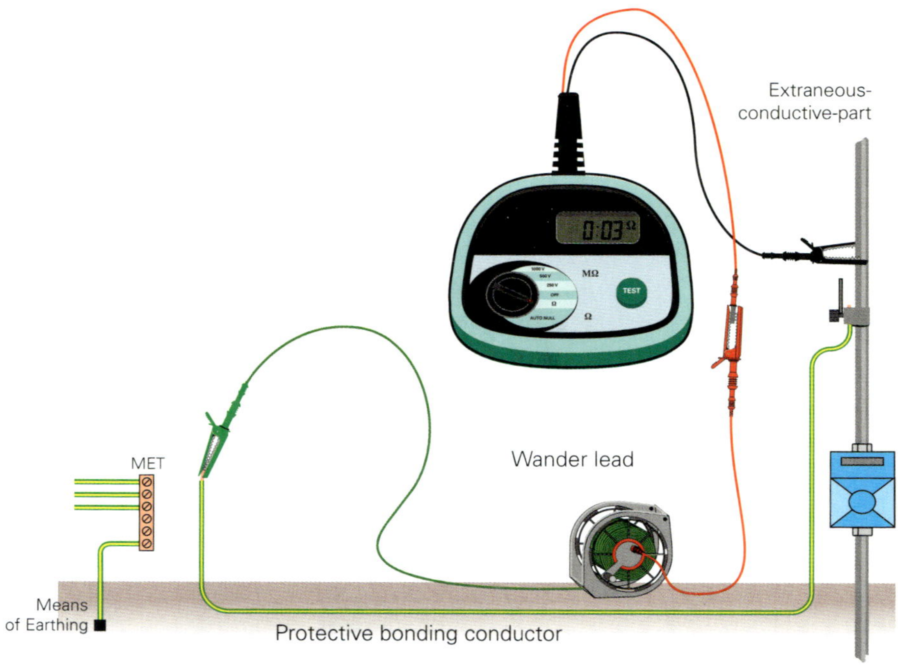

Fig 8.4.1 (i) Using the 'wander lead' method to measure the continuity of a main protective bonding conductor

Note: Where necessary paint or other coatings should be removed from metallic pipes to ensure a good electrical connection.

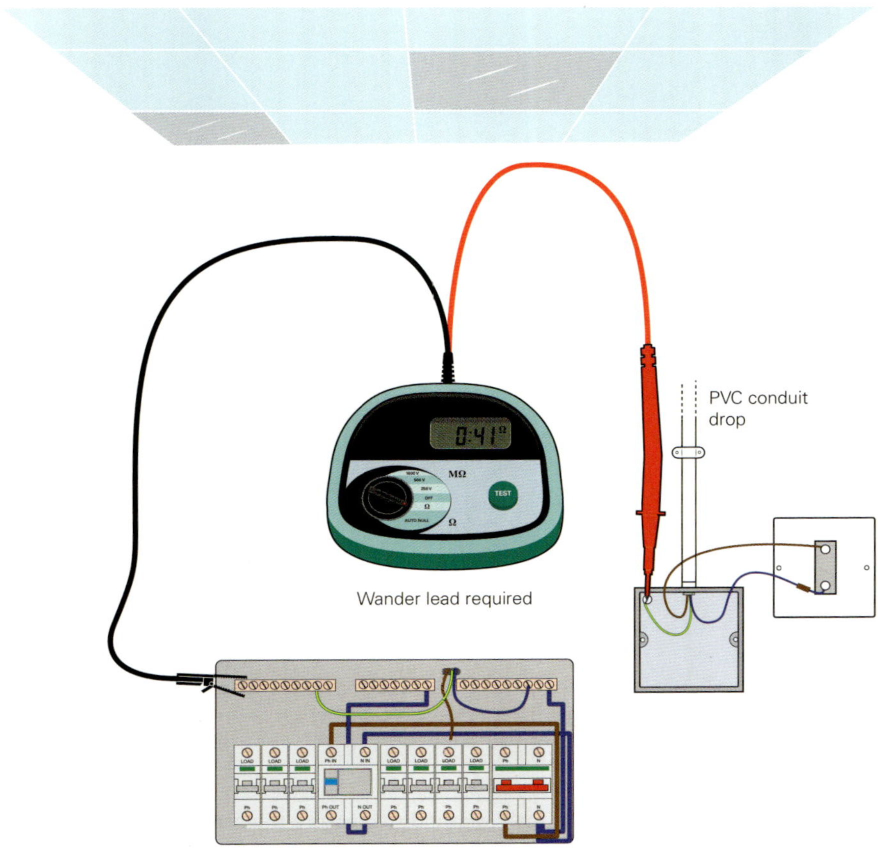

Fig 8.4.1 (ii) Wander lead test to verify continuity of a circuit protective conductor

Where a ferrous enclosure, such as steel conduit or steel trunking is used as a circuit protective conductor, the integrity of the enclosure should be verified for compliance with *BS 7671*. Where safe to do so, the continuity of the conduit or trunking may be determined using the wander lead method.

Fig 8.4.1 (iii) Continuity testing of a ferrous enclosure using the R_2 (wander lead) method

Note: Inspection of enclosures should confirm all joints provide durable electrical continuity and adequate mechanical strength.

The $(R_1 + R_2)$ method

This method is applicable to circuit protective conductors and their associated line conductors but not for main or supplementary bonding conductors. The test measures the series resistance of both the line conductor and the circuit protective conductor of a circuit. Long test leads are not required.

This 'dead' test may be carried out by temporarily connecting together the line conductor and circuit protective conductor at the supply end, (at the appropriate distribution board or consumer unit) and then connecting the instrument test leads to the terminals of the line conductor and the circuit protective conductor at each point and accessory on the circuit in turn.

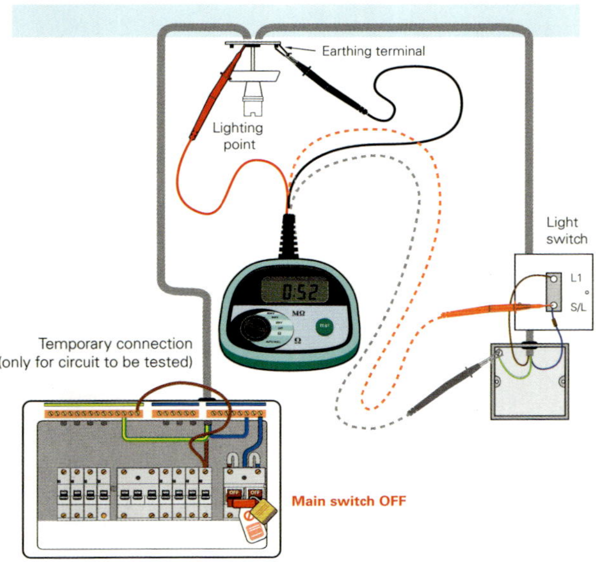

Fig 8.4.1 (iv) Continuity testing using the (R_1 + R_2) method

Test procedure:

- Securely isolate the supply (consumer unit) and where practicable disconnect any bonding connections that could affect the test readings.
- Make a temporary connection, at the supply end of the circuit (for example, at a distribution board/consumer unit) between the line and cpc of the circuit to be tested.
- Measure the resistance between line and cpc at each point and accessory on the circuit and, where appropriate, operate switches to confirm polarity.
- Record the reading at the electrically most remote point or accessory (the highest value) for the circuit (and confirm correct polarity for the circuit by inserting a 'tick').

Before the supply is reinstated:

Ensure any covers or enclosures that were removed for testing have been reinstated

- Re-connect any bonding conductors disconnected for the test.
- Remove the temporary connection.

The test readings at points and accessories on the circuit serve to verify that the protective conductor is continuous, and that the polarity is correct. The value of $(R_1 + R_2)$ measured at the point or accessory electrically most remote from the supply will be the maximum value, and it is this value that should be recorded in the appropriate row of the $(R_1 + R_2)$ column on the combined 'Schedule of Circuit details and Test Results'.

Note: The recorded value of $(R_1 + R_2)$ can be used to determine the earth fault loop impedance at the point or accessory at which the test is applied (refer to 'earth fault loop impedance' later in this Chapter).

8.4.2 Continuity of the live conductors of ring final circuits

The cables of a ring final circuit start and return to the same outgoing terminals of a consumer unit or distribution board. Therefore the live (line and neutral) conductors must form a complete unbroken loop without interconnections, as must the circuit protective conductor, unless it is formed by a continuous metal covering or enclosure (such as a steel conduit) containing all the live conductors.

A test instrument such as the type described earlier in this Chapter for testing the continuity of protective conductors should be used, and the following three-step procedure should be followed:

Step 1

Measure the end-to-end (loop) resistance of the

- line conductor (r_1)
- neutral conductor (r_n)
- cpc (r_2)

Where single-core cables have been used for the circuit wiring and the cpc has the same csa as the live conductors, the results of the circuit loop values obtained should be more or less equal (within ± 0.05 Ω, say). However, where the cpc has a reduced csa, such as in the case of 'twin and earth' type cables,the resistance of the cpc loop will be proportionally higher than the live conductors.

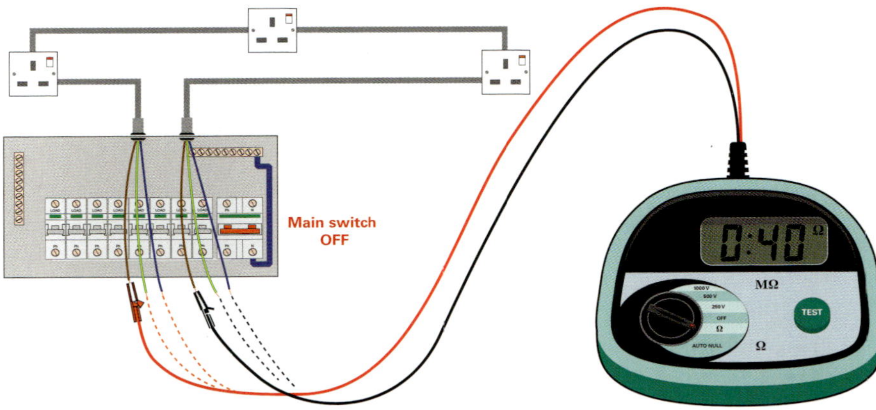

Line conductor mm^2	cpc (mm^2)	Expected ratio
2.5	1.5	1.67
4.0	1.5	2.67
6.0	2.5	2.4

Fig 8.4.2 (i) Measuring the ring final conductor loop values

As shown in the table, where the measured value of loop resistances for a line conductor of a 2.5 mm^2 flat twin and earth cable[1] *BS 6004* is 1 Ω, the measured loop resistance value for the cpc having a csa of 1.5 mm^2 will be (2.5 divided by 1.5) approximately 1.67 times higher than the line conductor loop (1.67 Ω).

The measured 'end-to-end' (loop) values should be recorded in the relevant boxes under 'Ring final circuits only' on the 'Schedule of Circuit Details and Test Results'. Where the cpc is formed by a metallic conduit or cable sheath, an end-to-end value will not be appropriate for the cpc. In such circumstances, N/A should be recorded for the end-to-end value and the resistance of the cpc should be obtained using 'R_2' (wander lead) method.

Step 2

Connect the outgoing line conductor of the circuit to the incoming neutral conductor of the circuit and vice versa at the distribution board or consumer unit, and measure the resistance between line and neutral terminals at every socket-outlet connected to the circuit.

[1] *BS 6004:2012 Electric cables – PVC insulated and PVC sheathed cables for voltages up to and including 300/500 V, for electric power and lighting*

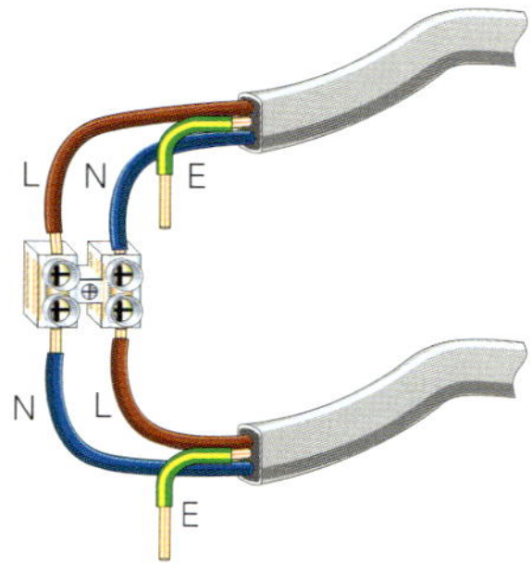

Fig 8.4.2 (ii) Cross-connection arrangement for Step 2

The results obtained should be substantially the same at each socket-outlet and should be approximately one quarter of the combined resistance of the line loop (r_1) and neutral loop (r_n) that is; $\frac{(r_1 + r_n)}{4}$

Note: Any socket-outlets connected as spurs will have a higher resistance value as a result of the resistance of the spur conductors.

Step 3

Next, the outgoing line conductor of the circuit should be connected to the incoming cpc of the circuit (and vice versa) at the distribution board or consumer unit and the resistance between the line and cpc terminals should be measured at every socket-outlet connected to the circuit under test.

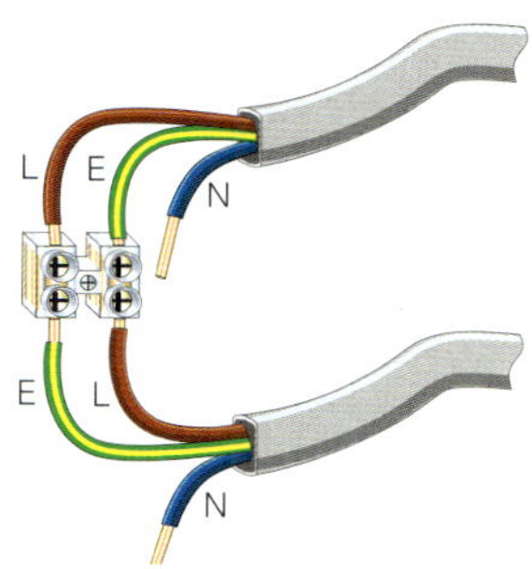

Fig 8.4.2 (iii) Cross-connection arrangement for Step 3

The results obtained should be substantially the same at each socket-outlet and should be approximately one quarter of the combined resistance obtained for the line loop (r_1) and cpc loop (r_2) that is; $\frac{(r_1 + r_2)}{4}$

Note: Any socket-outlets connected as spurs will have a higher resistance value as a result of the resistance of the spur conductors.

The highest value of resistance obtained should be recorded on the 'Schedule of Circuit Details and Test Results' as the $(R_1 + R_2)$ value of the circuit.

Where single-core cables have been used for the circuit wiring and the resistance readings obtained increase as the distance from the distribution board or consumer unit increases it is likely that the cross-connections have been made to conductors from the same leg of the circuit (i.e. both incoming or outgoing) and not to opposite legs. If this occurs, correctly cross-connect the circuit and start the testing again.

Note: When performing additions or alterations or periodic inspection, it should never be assumed that both legs of a ring final circuit have been connected into the same protective device. Having switched off the protective device, a voltage tester must be used to confirm that all live conductors are dead before disconnecting.

8.5 Insulation resistance

There is always a risk that cables might suffer damage during the installation process. For example, a cable installed in a partition could be damaged as a result of a misplaced fixing during the boarding of the partition, or cables pulled through conduit without due care may suffer insulation damage. For such reasons, Regulation 643.3.1 requires that the insulation resistance of installed circuits is measured between:

- live conductors, and
- between live conductors and the protective conductor connected to the earthing arrangement. (For the purpose of this test, the line and neutral conductors may be connected together, where appropriate.)

Insulation resistance testing should be carried out at appropriate times during the installation and on completion before the supply is connected. For example, after a cable has been concealed in a wall or partition, undertaking insulation resistance testing at that time will confirm whether the cable insulation has been damaged during the construction work. Testing at relevant points during the construction phase helps to ensure any damage that might have occurred is identified and rectified at an early stage of the project, rather than at the completion stage. Particularly as by that time the structure may, for example, be tiled, decorated and/or fitted with cabinets or units making the remedial work much more intrusive and costly.

For practical reasons it may be necessary or desirable to sub-divide the installation in order to carry out insulation resistance testing. Circuits may be tested collectively and wherever practicable a complete distribution board or consumer unit may be tested.

However, care must be exercised to ensure collective testing is appropriate especially where devices, such as RCDs / RCBOs are connected within the equipment. Where necessary circuits should be tested individually.

Table 8.1 Minimum insulation resistance test values

Circuit nominal voltage (V)	Test voltage DC (V)	Minimum insulation resistance (MΩ)
SELV and PELV	250	0.5
Up to and including 500 V with the exception of the above systems	500	1.0
Above 500 V	1000	1.0

The insulation resistance of the main switchboard and each distribution circuit tested separately with all its final circuits connected, but with current-using equipment disconnected, should have an insulation resistance not less than the appropriate value given in Table 8.1 (Table 64 of *BS 7671* refers).

Note: For the testing of fire detection and fire alarm systems in buildings refer to *BS 5839-1.*

The insulation resistance test of a functional extra-low voltage circuit should be performed at the same test voltage as that used for the primary side of the source of the FELV circuit.

8.5 (i) Test precautions

When insulation testing is carried out on completion of the electrical work precautions may be necessary to avoid damage to electronic devices, such as heating timers, controllers, lighting dimmer controls and the like. Equipment which might affect the test results and/or which might suffer damage from exposure to the test voltage being applied should be disconnected for the purpose of the test. Where it is not reasonably practicable to disconnect electronic devices, such as where a fixed low voltage socket-outlet incorporates an SPD, the person carrying out insulation resistance testing on low voltage circuits may adopt one of the following approaches:

- Perform the test with the live (line and neutral) conductors connected together and test between the live conductors and Earth only. (The term 'Earth' means that the protective conductor(s) are connected to the earthing arrangement.)
- Perform the insulation resistance test between conductors at a reduced test voltage of 250 V DC. However, where this option is used, the minimum acceptable insulation resistance remains 1 MΩ.

For the benefit of those who undertake future testing on the installation, whether as part of planned maintenance work or periodic testing, any circuit that has equipment connected to it which is vulnerable to damage from insulation resistance testing should be identified on the 'Schedule of Circuit Details and Test Results', and a label should be fixed to the appropriate distribution board or consumer unit (Regulation 514.9.1 refers).

8.5 (ii) The test procedure

Where it is practicable to test a complete distribution board or consumer unit the following procedure may be followed:

- Securely isolate the distribution board under test, including the neutral and disconnect all equipment that is vulnerable to damage or may affect the test result, such as electronic devices, LEDs and discharge lamps.
- All protective devices are either on or (in the case of fuses) in place.
- Perform test instrument checks to confirm the condition of the batteries, leads and contact probes are in sound condition. In particular, before selecting the insulation test range the leads should be checked for continuity to confirm that they are functioning correctly.
- Select the appropriate test voltage and (MΩ) range.
- **Live/Live:** Connect the instrument leads and test between the following:

For a single-phase installation test between:	Line to neutral
For a three-phase installation test between:	Line 1 to Line 2
(see Fig 8.5 (ii) a).	Line 2 to Line 3
	Line 3 to Line 1

(Then test the line conductors collectively or individually to the neutral.)

Measurement of the insulation resistance between one of the line conductors and neutral for a three-phase distribution board with its final circuits connected. Main neutral link open for the purpose of test.

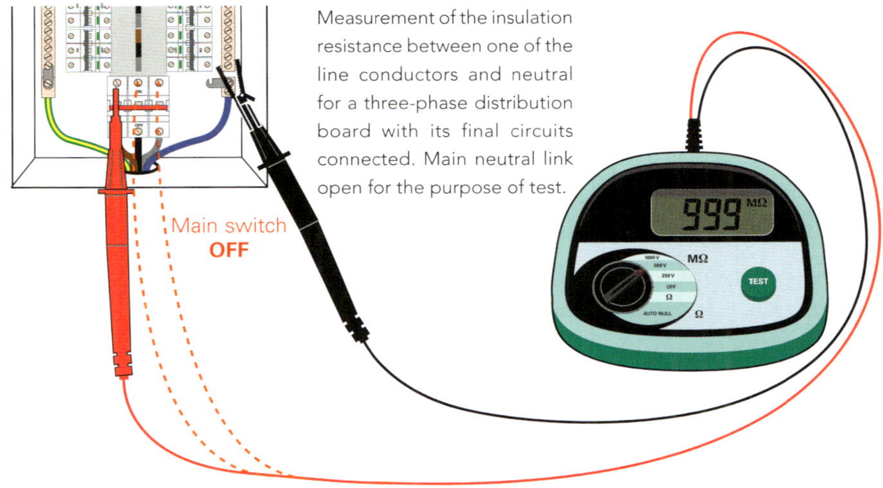

Fig 8.5 (ii) a. Insulation resistance between live conductors.

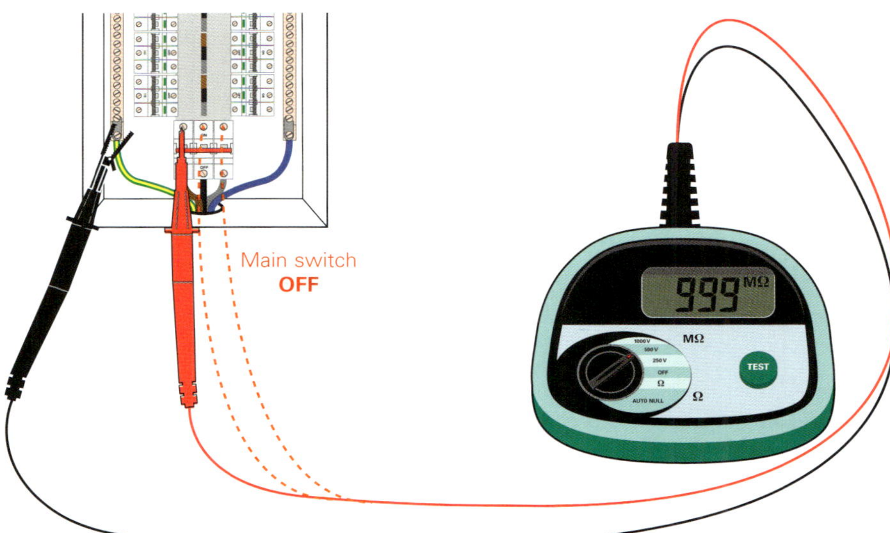

Fig 8.5 (ii) b. Insulation resistance between live conductors to Earth.

- **Live/Earth** Having taken the necessary precautions as before, the inspector should connect one lead of the test instrument to the earth terminal of the distribution board or consumer unit (ensuring the protective conductors are connected to the earthing arrangement) and the other lead should be connected to each live (line and neutral) conductor in turn. The insulation resistance of each live conductor to Earth should be measured and the lowest result (of each line to earth or line to neutral) recorded on the 'Schedule of Circuit Details and Test Results'. Where appropriate, the live conductors may be linked together for the purpose of the test.

If the whole distribution board or consumer cannot be tested, each circuit will need to be tested individually (On the load side of each protective device, in the 'OFF' position.) following the same procedure.

For initial verification insulation resistance readings should be high and values much higher than the minimum values of Table 64 of *BS 7671* should be obtained, generally readings of more than 200 MΩ would be expected for individual boards/circuits. Where the test readings of similar circuits show evident differences in insulation resistance further investigation should be undertaken to identify the reasons.

Before an addition or alteration is carried out to an existing installation the condition of the existing installation should be determined. Where insulation resistance testing of a distribution board or consumer unit reveals an insulation resistance lower than 2 MΩ, the circuits should be separated and tested individually in order to identify the cause of the low reading. The overall low reading may, depending on the particular wiring system, be due to an accumulation of individual circuit readings, and hence be considered acceptable. Alternatively, the low insulation reading may be due to one particular circuit, which would merit further investigation.

8.5 (iii) Protection by SELV, PELV or by electrical separation

Separated extra-low voltage (SELV), protective extra-low voltage (PELV) and electrical separation are protective measures against electric shock (refer to Chapter 6). In each of these protective measures, separation of live parts from other circuits and, except for PELV, separation from Earth is required.

To comply with the requirements of *BS 7671*, Regulation Group 643.4 requires that verification of the electrical separation conditions for each of these protective measures is confirmed by measurement of the insulation resistance, and the values obtained should be in accordance with Table 61 of *BS 7671*.

Test requirements for SELV circuits

As required by Regulation 643.4.1, insulation resistance testing of SELV circuits is required to confirm that the live parts of the SELV circuit satisfy the requirements of Section 414 of *BS 7671* for the provision of:

- basic insulation between live parts and other SELV (or PELV) circuits,
- basic insulation between live parts and Earth, and
- protective separation from live parts of any other systems by means of double or reinforced insulation, or basic insulation and protective screening for the highest voltage present.

The source for a SELV system must be of a type listed in Regulation 414.3. A safety isolating transformer complying with *BS EN 61558-2-6* or *BS EN 61558-2-8* is commonly used. These types of transformer are designed to provide protective separation (equivalent to double or reinforced insulation) between the primary and secondary windings, and have no provision for an earth connection on the secondary ELV side.

The insulation resistance test values obtained for SELV circuits should satisfy Table 64 of *BS 7671*, based on the highest voltage circuit present. The test requirements for SELV circuits are summarised in the following Table.

Table 8.2 Tests requirements for SELV circuits

Tests requirements for SELV circuits	Insulation resistance test voltage	Minimum acceptable value (from Table 61)
Test between all live conductors of the SELV* circuit, and Earth and, where present, between the live conductors of each SELV circuit and the conductors of any other SELV (or PELV) circuit.	250 V DC	0.5 MΩ
Test between all live conductors of the SELV* circuit, and any live parts of a low voltage (or FELV) circuit. If the nominal circuit voltage is greater than 500 V the insulation resistance test should be carried out at 1000 V DC (Regulation 643.4 refers).	500 V DC or 1000 V DC	1 MΩ

* Where appropriate the live conductors may be connected together for the purpose of the test.

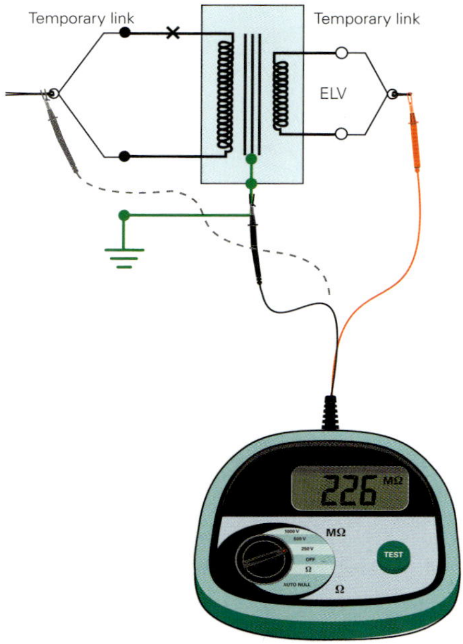

Fig 8.5 (iii) (a) Test for separation between the SELV circuit and Earth

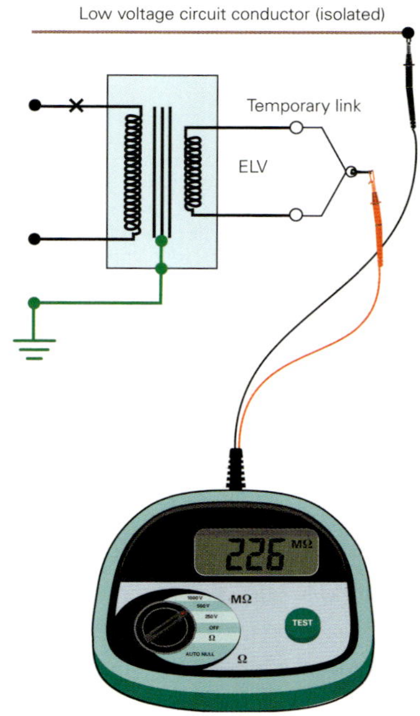

Fig 8.5 (iii) (b) Test between the SELV circuit and any live parts (isolated) in close proximity

Test requirements for PELV circuits

A PELV system is similar to a SELV system, except that the extra-low voltage circuit is generally connected to Earth at one (or more) point(s).Therefore the test requirements for PELV are the same as SELV, except that it is not necessary to perform an insulation resistance test between the PELV circuit and Earth (Regulation 643.4.2 refers).

8.5 (iv) Test requirements for electrical separation

Electrical separation is a protective measure that may be applied to circuits operating at a voltage of up to 500 V, and should not be confused with SELV or PELV.

The source for electrical separation must provide at least simple separation at a voltage not exceeding 500 V (Regulation 413.3.2 refers), and should have no connection of live parts of the separated circuit to another circuit or to Earth or to a protective conductor (Regulation 413.3.3 refers). To confirm that separation is achieved in accordance with Section 414 of *BS 7671*, insulation resistance testing should be undertaken and the values obtained should comply with Table 64 of *BS 7671* (Regulation 643.4.3 refers).

Where electrical separation is used to supply more than one item of equipment the requirements of Regulation 643.4.3 for disconnection under fault conditions must be satisfied.

Insulation resistance/impedance of floors and walls

Where the protective measure 'Non-conducting location' is used, such as in certain locations within hospitals or test laboratories, compliance with the requirements of Regulation Group 418.1 is required. For such purposes, insulation resistance testing should be applied in accordance with Regulation 643.5.1. Guidance on the testing is given Appendix 13 of *BS 7671*.

8.6 Polarity

Incorrect polarity can give rise to danger in a number of ways; for example, parts of the installation may remain connected to the line conductor when switched off by a single-pole device and in the event of an overload, the circuit-breaker or fuse protecting that part of the installation would disconnect the neutral of the circuit, leaving the load at full line voltage. For such reasons, Regulation 643.6 requires polarity testing is carried out to verify:

- all fuses, single-pole control devices and single-pole protective devices are connected in the line conductor only, and
- in circuits having an earthed neutral conductor, edison screw and other centre-contact lampholders (except for E14 and E27 lampholders to *BS EN 60238*) should be connected in such a way that the line conductor is connected to the centre contact, and the neutral to the screw thread or the part of the lampholder which makes contact with the lamp cap, and
- wiring has been correctly connected throughout the installation.

Polarity testing is carried out before the circuit is energised (dead polarity testing) using a low-reading ohmmeter or continuity test instrument. Much of the polarity testing can be carried out during the process of testing cpc continuity by using the ($R_1 + R_2$) test method described earlier in this Chapter (see 'Continuity of conductors').

8.7 Installation earth electrode resistance, (R_A)

Where an earth electrode provides the means of earthing for an installation, the electrode resistance to Earth (R_A) must be measured to verify that it provides a satisfactory means of earthing. Where a measurement of R_A is not practicable, the measured value of external earth fault loop impedance may be used as an approximate value (Regulation 643.7.2 refers).

The measured value of the electrode's resistance to Earth should be recorded in the appropriate section of the certificate or report along with the type of electrode and its location.

Suitable types of earth electrodes are listed in Regulation 542.2.2, and information on the installation of particular types of earth electrodes is given in *BS 7430*[1]. However, regardless of the type of electrode installed, the magnitude of the measured value of earth resistance must be sufficiently low enough to confirm that the connection will function reliably. As stated in 'Note 2' of Table 41.5 of *BS 7671* a value greater than 200 Ω may not be reliable. One method of measuring the resistance of an earth electrode using a proprietary tester is described as follows.

8.7 (i) Measurement of installation earth electrode resistance using a proprietary tester

Earth electrode resistance may be measured using a proprietary test instrument in accordance with the instrument manufacturer's instructions, using temporary electrodes placed in linear or triangular formations.

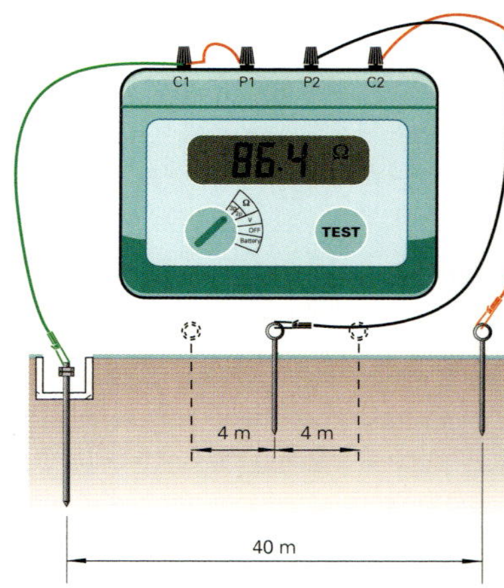

Fig 8.7 (i) Measurement of installation earth electrode resistance using a proprietary tester

Note: The ground adjacent to the installation earth electrode must not be treated with water or brine, as this would make the resistance measurement invalid for the purpose of verification or reporting.

[1] *BS 7430:2011+A1:2015 Code of practice for protective earthing of electrical installations*

In the case of the four-terminal instrument shown in Fig 8.7 (i), the instrument passes a current through the ground between the two current terminals C1 and C2. Simultaneously, the voltage drop between the installation earth electrode and the general mass of Earth is determined by means of the voltage (potential difference) at terminals P1 and P2.

If the test instrument has to be situated some considerable distance from the electrode under test, two separate leads should be run out from C1 and P1 to minimise the effect of lead resistance, and the link shown should be formed at the electrode under test.

Test procedure

With the installation isolated and the earth electrode disconnected from the earthing conductor the following procedure may be adopted.

- The temporary (current) electrode C2 is placed in the ground at sufficient distance away from the installation earth electrode under test so that the resistance areas of the two electrodes do not overlap. Unless the soil resistivity is very high, a distance of 40 metres is usually sufficient.
- A second temporary (potential) electrode P2 is placed approximately mid-way between these two electrodes (approximately 20 m from each).
- The test is conducted and the instrument should calculate and display an ohmic value for the resistance of the installation earth electrode under test.
- One reading is taken with P2 in the mid-way position, and a further two readings are taken with P2 moved either side of the mid-way point. The distance P2 is moved being approximately 10 % of the distance between C2 and the electrode under test.

Where the three readings are substantially the same, the mean (average) of the three readings is taken as the resistance of the earth electrode. Where the readings are markedly different, the electrode C2 should be moved further away from the installation earth electrode, and the test procedure repeated. When three acceptable resistance readings have been obtained, the mean value of the three readings should be recorded as the 'Electrode resistance to Earth' on the certificate or report.

Note: A sufficient reduction in the resistance of the temporary electrodes may be achieved by driving longer temporary rods into the ground.

It should be noted that there are instruments available that will allow the earth electrode to be tested without having to disconnect it from the installation. Such instruments, which are particularly beneficial for periodic testing, have a measurement mode which uses an external current clamp to measure the current through the earth electrode. Where such instruments are used the test should be performed in accordance with the particular instrument manufacturer's instructions.

8.7 (ii) Measuring R_A using an earth fault loop impedance test instrument

As stated earlier, where a measured reading of R_A is not practicable, the measured value of external earth fault loop impedance (Z_e) for the installation may be used as the approximate value of earth electrode resistance to Earth.

External earth fault loop impedance, Z_e is a 'live test' so the test method is described later in this Chapter see 'Measurement of external earth fault loop impedance, Z_e '.

It should be noted that the test result of the Z_e test is the impedance of the whole earth fault loop path external to the installation but as the impedance of the other parts of the earth fault loop can be expected to be relatively low, the reading obtained is taken as the approximate value of the resistance of the installation earth electrode.

Therefore, the measured value of Z_e may also be recorded in the appropriate section of the certificate / report as the value of the 'Electrode resistance to Earth'.

8.7 (iii) Verification of earth electrode test results

Where the protective measure is ADS, as is normally the case, it is essential that the resistance of the installation earth electrode is determined in order to verify that the installation circuits satisfy the disconnection requirements of *BS 7671*.

BS 7671 recognises the use of an RCD or an overcurrent protective device to provide automatic disconnection of a circuit forming part of a TT system under fault conditions (Regulation 411.5.2 refers). However an RCD is preferred as the latter may only be used where a suitably low value of earth fault loop impedance (Z_s) is permanently and reliably assured.

Where an RCD is used for earth fault protection, Regulation 411.5.3 requires the following conditions to be fulfilled:

- The relevant disconnection time of Regulation 411.3.2.2 or 411.3.2.4 should be satisfied, and
- $R_A \times I_{\Delta n} \leq 50\ \text{V}$

Where:

$\boldsymbol{R_A}$ is the sum of the resistance of the earth electrode and protective conductor connecting it to the exposed-conductive-parts (in ohms), and $\boldsymbol{I_{\Delta n}}$ is the rated residual operating current of the RCD.

These requirements are met if the Z_s of the final circuit protected by the RCD meets the requirements of Table 41.5 of *BS 7671*. Where an overcurrent protective device is used, the conditions specified in Regulation 411.5.4 should be fulfilled.

Note: Where an RCD is used for fault protection the circuit should also incorporate an overcurrent protective device in accordance with Chapter 43.

For details on the disconnection of distribution and final circuits refer to 'Measurment of earth fault loop impedance, (Z_S)'.

Live testing

8.8 Polarity of incoming supply

Before the circuits of an installation are energised, the polarity of the incoming supply should be confirmed. The specific procedures required to verify the polarity of a live supply are dependent on the type of supply system (TT, TN-S or TN-C-S) and are summarised in the following tables.

TT and TN-S systems

Using a GS38 compliant test lamp or voltmeter the procedure shown in Table 8.1 may be used to verify the supply polarity for a TT or TN-S system.

Table 8.3 Live polarity testing procedure applicable for TT and TN-S supply systems

Connections of voltage indicator leads	Required outcome
Line to neutral	Lamp (or 230 V indicator) should light
Neutral to earth	Lamp (or 230 V indicator) should NOT light
Line to earth	Lamp (or 230 V indicator) should light
Connections of voltage indicator leads	**Required outcome**
L1 to L2	Lamp should be fully bright or 400 V indicator should light
L1 to L3	
L2 to L3	
L1 to N	Lamp should light or 230 V indicator should light
L2 to N	
L3 to N	
L1 to E	Lamp should light or 230 V indicator should light
L2 to E	
L3 to E	
N to E	Lamp or 230 V indicator should NOT light

TN-C-S system

It should be noted that due to the combined neutral and earth connection forming part of the TN-C-S system, the sole use of a two-pole voltage indicator is not suitable to identify incorrect supply polarity. However, supply polarity for a TN-C-S system may be verified by using a single-pole voltage indicator in conjunction with the two-pole voltage detector using the following procedures. All instruments used should comply with GS38.

Table 8.4 Live polarity testing procedures applicable for TN-C-S supply systems

Using a suitable single-pole voltage indicator	
Contact to line	Lamp should light
Contact to neutral	Lamp should NOT light
Contact to line	Lamp should light (confirming continued correct operation of indicator)
Connections of voltage indicator leads	**Required outcome**
Line to neutral	Lamp (or 230 V indicator) should light
Neutral to earth	Lamp (or 230 V indicator) should NOT light
Line to earth	Lamp (or 230 V indicator) should light

8.9 Protection by automatic disconnection of supply (ADS)

ADS, is the most common method used to provide fault protection for circuits forming part of TN or TT systems, and may be achieved by either of the following devices:

- An overcurrent protective device
- An RCD

To verify that devices installed for fault protection will provide ADS in accordance with *BS 7671*, the earth fault loop impedance path, Z_S, should be determined for every distribution circuit and final circuit that relies on ADS as its protective measure against electric shock (Regulation 643.7 refers).

Where ADS is provided by an overcurrent protective device, it must be verified that the maximum earth fault loop impedance permitted for the device, to achieve the required disconnection time, is not exceeded. (Measurement of earth fault loop impedance, (Z_S) is covered later in this Chapter.)

Where an RCD is used to provide fault protection the maximum values of Z_S given in Table 41.5 are applicable. Guidance on the verification of RCDs used to provide fault protection is covered later in this Chapter under 'Operation of RCDs'.

It should be noted that irrespective of whether an installation forms part of a TN or TT system, where an RCD is used to provide fault protection the circuit should also incorporate an overcurrent protective device in accordance with Chapter 41 of *BS 7671* (Regulations 411.4.204 and 411.5.2 refer).

As part of the verification for ADS, the means of earthing for the installation should be confirmed before the installation is energised. For this purpose the external earth fault loop impedance Z_e for an installation is required to be determined, by enquiry, measurement or calculation, and recorded on the the certificate or report.

8.10 Measurement of external earth fault loop impedance (Z_e)

The external fault loop impedance of an installation Z_e forms part of the earth fault loop impedance path for each circuit that relies on Automatic Disconnection of Supply (ADS) as the protective measure. Although Z_e may be determined by enquiry, a measurement of Z_e should still be undertaken as part of the initial verification to confirm that the intended means of earthing is present and its impedance is suitable for the type of earthing arrangement. A simplified diagram of the external earth fault path for a single-phase supply, is shown in Fig 8.10.

The test is performed between the line conductor and the earthing conductor, as shown. The external earth path will depend on the type of supply, for a TN-C-S supply the earth return path will be via the supply neutral conductor whereas for the TT system shown in the example the earth return path is through the general mass of Earth.

8.10 (i) Conducting an external earth fault loop impedance test

For the purpose of the test it is necessary to disconnect the earthing conductor from the main earthing terminal or otherwise separate it from all parallel earth paths such as bonding and circuit protective conductors. Any parallel earth paths present during the test would result in a false reading of Z_e, being obtained and might perhaps conceal the fact that the intended means of earthing is defective, or even non-existent.

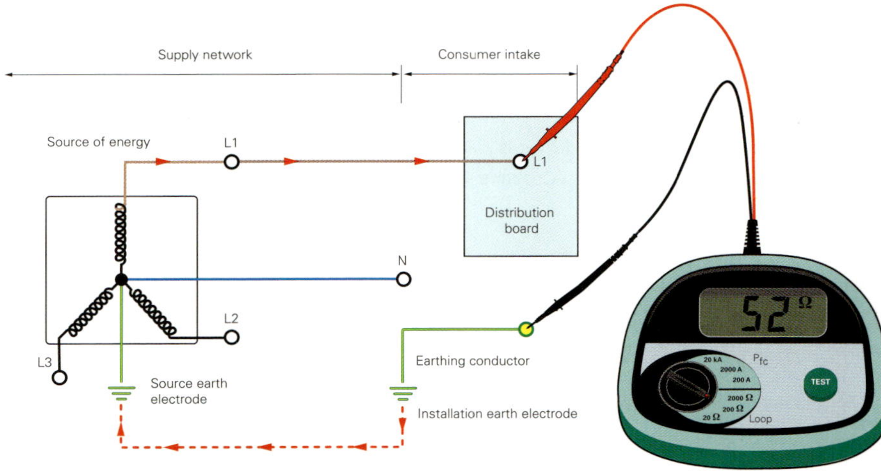

Fig 8.10 The external earth fault loop impedance path for a TT single-phase supply system

Where the supply is provided with an earth connection from the electricity distributor's network, such as from a TN-S or TN-C-S system, it is unlikely that Z_e will exceed one ohm, so the lowest range of the instrument will be the most appropriate. Typical maximum values of Z_e for single-phase supplies of up to 100 A published by electricity distributors are:

- TN-S 0.80 Ω
- TN-C-S 0.35 Ω

Whilst for a TT system a value of 21 Ω is declared, this value is based on the supplier's source earthing electrode and does not include the impedance of the consumer's earth electrode(s). For an installation connected to a TT system the earth fault return path from the consumer's installation to the supply transformer is via the general mass of earth. Therefore, the measured value of Z_e is likely to fall within the higher resistance range of the test instrument, see 'Installation Earth Electrode Resistance, (R_A)'.

8.10 (ii) Measurement of Z_e at a single-phase board

With the installation isolated and the earthing conductor disconnected from the MET, the test is performed at the origin of the installation between the incoming line conductor and the earthing conductor, as shown in Fig 8.10 (ii) a.

For a TT earthing arrangement the measured value of Z_e may be taken as an approximate value for the earth electrode resistance R_A (at the origin of the installation) with the installation disconnected from the supply. Fig 8.10 (ii) b, shows a simplified diagram of the test conducted on one phase of a three-phase installation which is connected to a TT system. With the main distribution isolated and the earthing conductor disconnected from the MET, the test is performed between the line conductor and the earthing conductor as shown. For a three-phase supply the test must be repeated for each phase and the highest reading should be recorded on the certificate.

Before disconnecting the earthing conductor (or any main protective bonding conductors) the installation should be isolated from the supply.

Note: The test should be performed in accordance with the instrument manufacturer's instructions and should only be conducted when the instrument indicates that it is safe to do so.

Test procedure:

- Check that the test instrument, leads, probes and crocodile clips (if any) are suitable for the purpose and in good serviceable condition
- Open the main switch to disconnect the whole of the installation from the source of supply, and secure the switch in the open position
- Disconnect the earthing conductor from the main earthing terminal or otherwise separate it from all main bonding conductors and circuit protective conductors
- Observing all precautions for safety, apply the test probes (see Fig 8.10 (ii))
- Check the polarity indicator for correct connection
- Press the test button
- For a three-phase supply, test each of the remaining phases to Earth, Fig 8.10 (ii) b.

On completion of the test and before the supply is restored to the installation the earthing conductor or other protective conductors that were disconnected to facilitate the measurement of Z_e must be reconnected.

The highest test reading should be recorded in the relevant part of the certificate or report.

Note: Where an installation is in service, such as in the case of periodic inspection, the incoming supply to the building must not be isolated unless prior agreement has been obtained and those that will be affected by the disconnection have been given prior notice.

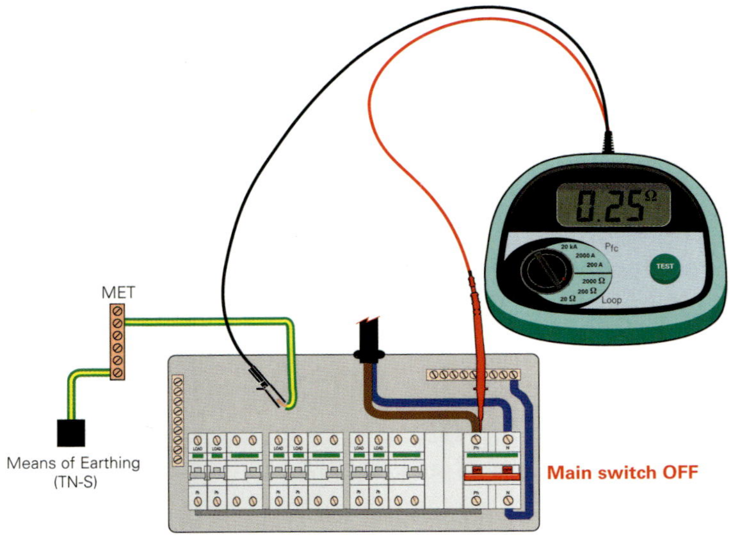

Fig 8.10 (ii) a. Measurement of Z_e at a single-phase consumer unit

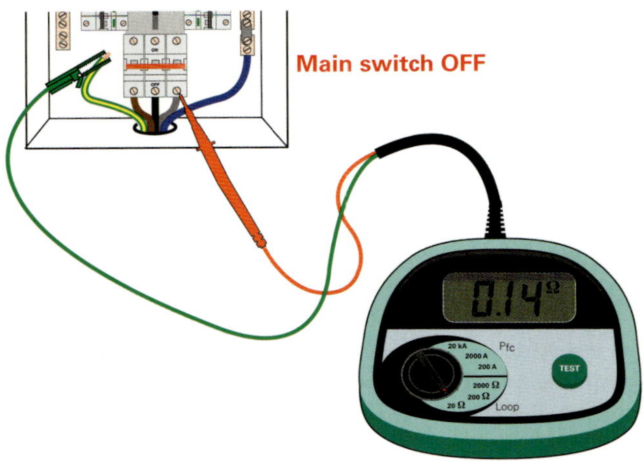

Fig 8.10 (ii) b. Measurement of Z_e at a three-phase distribution board

8.11 Measurement of maximum prospective fault current

Regulation 643.7.3.201, requires prospective fault current to be determined at the origin and other relevant points within the installation. Where the maximum prospective fault current is determined by testing, the test should be performed at the origin of the installation where the fault current is highest, such as at the main switch or other switchgear supplied directly from the distributor's equipment. For these reasons, where a test is conducted to obtain the maximum prospective fault current for an installation, it is normally performed directly after external impedance testing described previously.

The earthing conductor, main protective bonding conductors and circuit protective conductors should all be connected during these tests, because the presence of these and any other parallel paths to Earth may reduce the impedance of the fault loop and so increase the prospective fault current.

Note: Prospective fault current tests should be carried out on the output terminals of a suitably rated protective device, and if such a device is not installed a temporary one should be fitted. Irrespective of whether fused test leads are used, the test should only be made where overcurrent protection is provided between the point of connection of the instrument leads and the transformer (Appendix 14 of *BS 7671* refers).

The purpose of the test is to confirm that the fault protective devices installed are capable of safely interrupting the highest fault currents that may occur within an installation. In general the short-circuit capacity of the protective device should be no less than the maximum prospective fault current at the point in the circuit at which it is installed (Regulation 434.5.1 refers). The measured values should be compared against the relevant installation design criteria to confirm the fault current rating of the installed equipment satisfies the requirements of *BS 7671* (Regulation 643.1 refers).

For NICEIC and ELECSA certificates and reports the maximum prospective fault current (I_{pf}) must be recorded in the relevant part of the section entitled 'Supply characteristics and earthing arrangements'.

8.11 (i) Conducting prospective fault current tests

Single-phase installations

Generally, testing to determine the prospective fault current at the origin or elsewhere within a domestic installation is unnecessary. A consumer unit complying with *BS EN 61439-3* is considered to be adequate to safely withstand a prospective fault current of up to 16 kA (Regulation 536.4.201 refers), and as a result of this conditional rating, applying tests to determine prospective fault current at either the origin or for final circuits would, in many cases, be unnecessary. In such circumstances, the maximum prospective fault current should be recorded as I6 kA.

Live testing should be avoided where the information required can be obtained by another method, such as enquiry. Where measurement is required in a single-phase installation, the prospective fault current I_{pf} is the greater of either the fault current between:

- the line conductor and neutral conductor (prospective short-circuit current), or
- the line conductor and Earth (prospective earth fault current).

The highest reading obtained from these tests should be recorded as the value of I_{pf}.

The prospective earth fault current at the origin can be calculated using, for example, the measured loop impedance value Z_e.

$$I_{pf} = \frac{V}{Z_e}$$

General test Procedure:

- Turn main switch 'OFF'.
- Check that the test instrument, leads, probes and crocodile clips (if any) are suitable for the purpose, and in good serviceable condition.
- Select the appropriate range and scale (for example, for a TN system earthing arrangement, prospective fault current 20 kA). Where necessary the range will need to be adjusted according to the particular test. For example, due to the high impedance of the earth fault path of a TT supply system, the prospective earth fault current is normally very low compared with the short-circuit fault current.
- Observing ALL precautions for safety, connect the instrument to the incoming energised supply to measure a line to neutral value, as shown in Fig 8.11 (i) a.
- Check the polarity indicator (if any) on the instrument for correct connection.
- Operate test button and record the reading.
- Repeat the above steps, as appropriate, with the instrument connected for measurement of the line to earth value, as shown in Fig 8.11 (i) b.

For installations having more than one supply source, such as an installation connected to a standby generator, the higher value of prospective fault current taking into account all sources should be recorded on the certificate. Generally, for additional supply sources such as a standby generator or a microgenerator, prospective fault current testing is not practicable. Information on the maximum output current of these additional supply sources should be obtained from the supply source manufacturer's documentation.

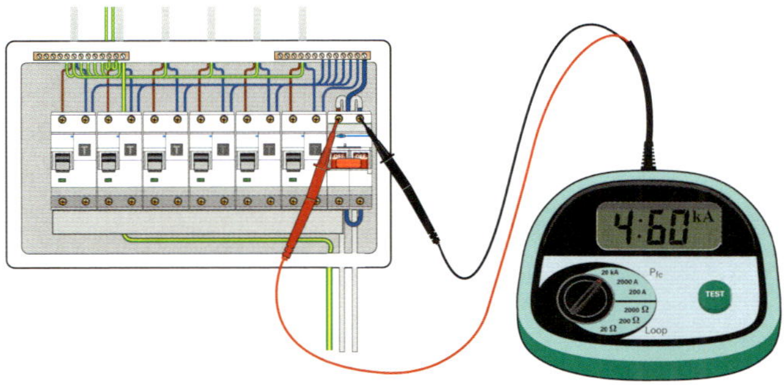

Fig 8.11 (i) a. Measurement of prospective short-circuit current

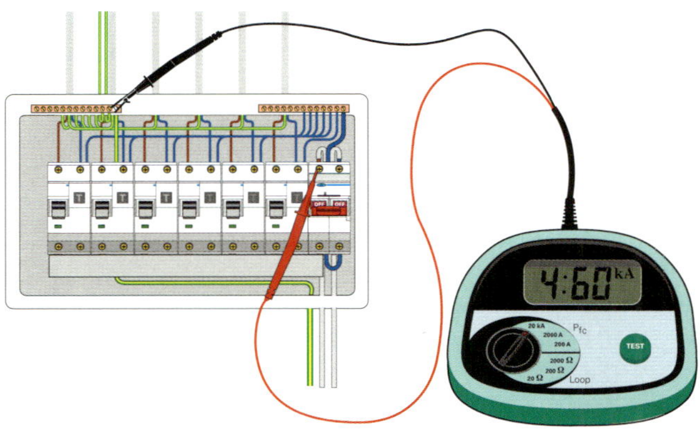

Fig 8.11 (i) b. Measurement of prospective earth fault current

8.11 (ii) Three-phase installations

In a three-phase installation the highest prospective fault current is the result of a simultaneous short-circuit fault between all line conductors (symmetrical fault). An approximation of this fault is determined by measuring the single-phase short-circuit current described earlier, and repeating the test for each phase. The highest reading obtained should be multiplied by a factor of two and then recorded as the maximum prospective fault current in the relevant part of the certificate or report (Appendix 14 of *BS 7671* refers).

For example, if the highest single-phase (line-to-neutral) prospective fault current measurement is 4.6 kA, as shown in Fig 8.11 (ii), the maximum three-phase prospective fault current should be recorded as 4.6 kA x 2 = 9.2 kA.

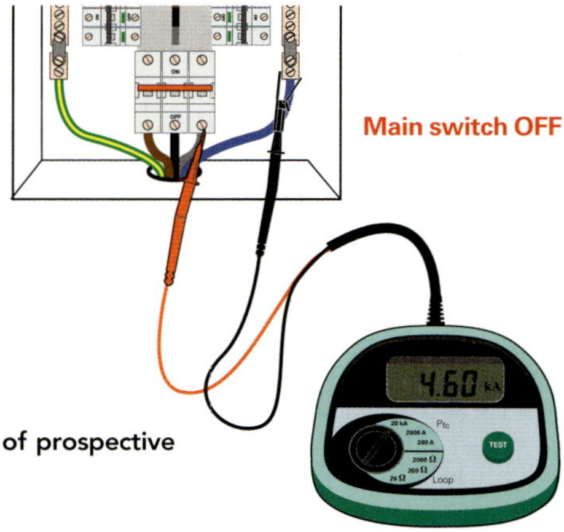

Fig 8.11 (ii) Measurement of prospective short-circuit fault current

Note: The image shows a two-lead instrument connected, but three-lead test instruments are also available so in each case refer to the particular instrument manufacturer's instructions.

Safety Warning: An instrument rated at 230 V must not be connected between two line conductors

8.11 (iii) Maximum short-circuit breaking capacities of protective devices.

Circuit-breakers

The rated short-circuit capacity (I_{cn}) for a circuit-breaker manufactured to the *BS EN 60898* series or RCBO manufactured to the *BS EN 61009* series is marked on the device, and is the maximum fault current, in amperes, that the device can safely interrupt. The value is assigned by the manufacturer on the basis of type-tests. The device is designed to break a fault of this magnitude only once.

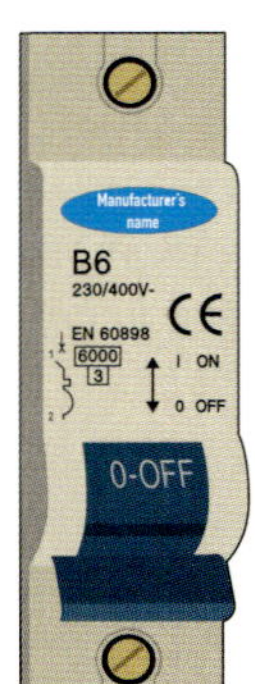

Fig 8.11 (iii) Breaking capacities of protective devices

The maximum fault current that the device is designed to break more than once, is called the 'service short-circuit capacity' (I_{CS}). This may be less than the ultimate short-circuit capacity marked on the device, so where necessary the manufacturer's data should be consulted for details.

Fuses

Short- circuit capacities for fuses are not so straightforward and as a guide typical values are:

- *BS 3036* - 1 kA to 4 kA (depending on duty rating)
- *BS 1361* - 16.5 kA
- *BS 88-2* - 80 kA (AC) 40 kA (DC)
- *BS 88-3* - 16 kA (Type I) 31.5 kA (Type II)

The manufacturer's data for the particular device should be consulted.

Propective fault current testing should, where necessary, be carried out at other relevant points in a circuit, particularly at the furthest points, to confirm that the fault current is sufficient to cause operation of the protective device before the permitted limiting temperature of any conductor or cable is exceeded (Regulation 434.5.2 refers).

8.12 Measurement of Earth fault loop impedance (Z_S)

Measured values of Z_S can be obtained wholly by means of an earth fault loop impedance test, where it is safe to do so. The purpose of the test is to verify that the protective devices installed for distribution and final circuits will operate within the time required by *BS 7671*.

As shown, measurements of Z_S should be made at the furthest point of every distribution circuit and final circuit, but before earth loop impedance measurements are undertaken the continuity of protective conductors should be verified (Regulation 643.7.2 refers).

Alternatively, the (R_1 + R_2) test results obtained during continuity testing may be used, in conjunction with test results obtained by the use of the loop impedance test instrument for other parts of the installation to provide calculated values of Z_S.

Note: For the purposes of initial verification appropriate drawings and information should be made available so that the furthest point of each circuit may be identified easily. However, this may not be the case for periodic inspection, so where necessary measurements of Z_S may need to be taken at each point and accessory of the circuit in order to determine the furthest point.

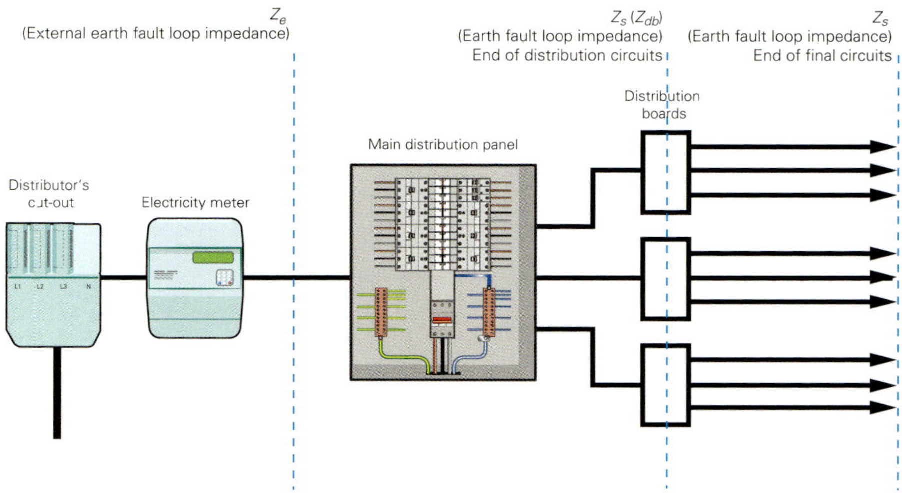

Fig 8.12 Positions where measurements of earth fault loop impedance may be required

8.12 (i) The earth fault loop impedance path

A simplified diagram of the earth fault loop path Z_s is shown. If the impedance of the fault path indicated is too high the fault current will be insufficient to operate the protective device within the required time. The complete earth fault loop path is comprised of the external impedance Z_e and the impedance within the installation ($R_1 + R_2$).

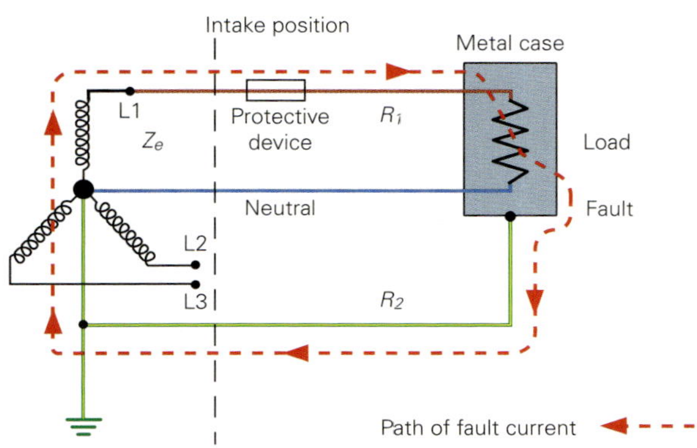

Fig 8.12 (i) The earth fault loop impedance path for a typical TN-S system

8.12 (ii) The Procedure for measuring Z_S with a loop impedance test instrument

This is a 'live' test so all necessary precautions must be taken to prevent danger when conducting the test. Testing should only be carried out after a thorough inspection and continuity testing have confirmed, so far as is reasonably practicable, that all circuit protective conductors, and main and supplementary bonding conductors, are in place.

Prior to performing the test, the test instrument and its connections including leads, clips and probes, should be checked to confirm they are in good condition, and before the test button is operated it should be confirmed that the instrument LEDs indicate that it is safe to proceed with the test.

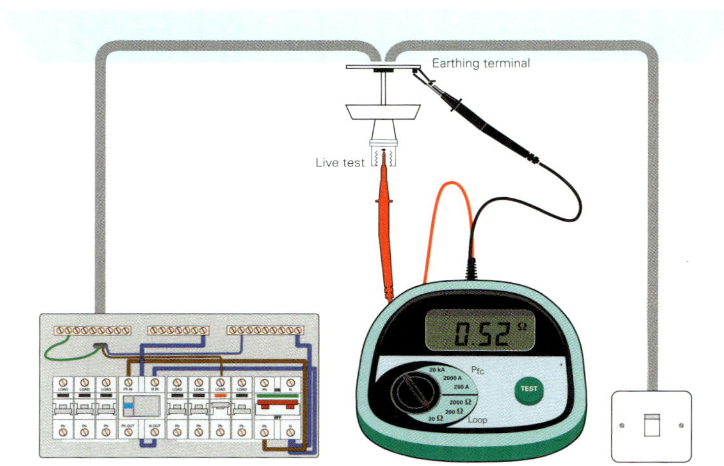

Fig 8.12 (ii) a. Earth loop impedance test on a lighting circuit

For lighting circuits, connect one of the instrument leads to the earth terminal before connecting the second lead to the line or switch line terminal at the lighting enclosure or ceiling rose at the electrically furthest point or accessory. Perform the test and then switch off the circuit on completion of the testing. Ensure all covers removed for the purpose of testing are properly reinstated.

For testing socket-outlets, a proprietary test lead with a fitted plug should be used. As a minimum, testing should be performed on the outlet which is situated electrically furthest from the origin of the circuit. Where this is not known, test all socket-outlets connected to the circuit and record the highest value on the test schedule.

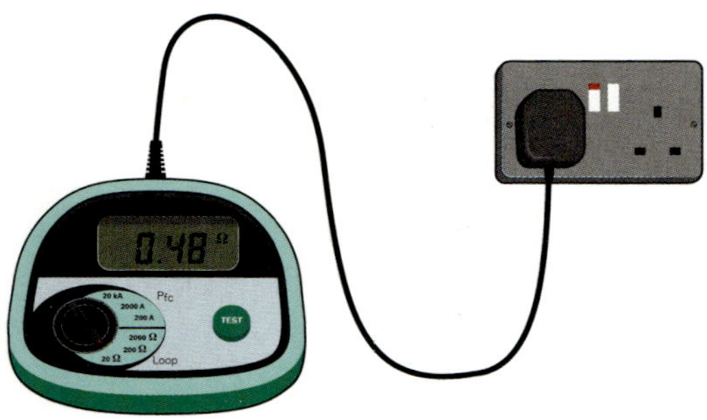

Fig 8.12 (ii) b. Z_S test at a socket-outlet using a proprietary test lead with a fitted plug

8.12 (iii) Determination of Z_s using $(R_1 + R_2)$ measured values

As previously indicated, an $(R_1 + R_2)$ test result for a given point in a circuit (generally the point which is electrically furthest from the supply) can be used to determine a value for the earth fault loop impedance, Z_s, at that point. To do this, it is also necessary to know the value of earth fault loop impedance, Z_{db}, at the distribution board or consumer unit or control panel directly supplying the circuit.

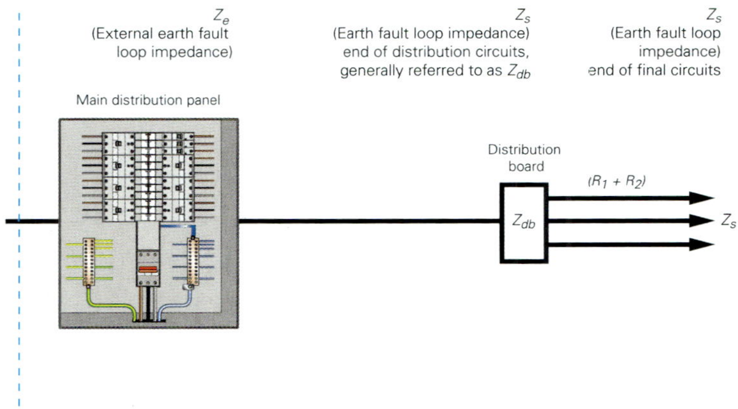

Fig 8.12 (iii) Determination of Z_s using $(R_1 + R_2)$ measured values

As shown, the earth fault loop impedance for the final circuit is determined by adding the measured value of Z_s taken at the distribution board (Z_{db}) and the circuit continuity reading $(R_1 + R_2)$ using the following formula:

$$Z_s = Z_{db} + (R_1 + R_2)$$

where:

Z_s is the earth fault loop impedance (Ω)

Z_{db} is the earth fault loop impedance measured at the distribution board to which the final circuit is connected

R_1 is the resistance of line conductor

R_2 is the resistance of protective conductor

Where the distribution board is at the origin of the installation, circuit impedance may be determined using the external earth fault loop impedance Z_e, as follows:

$$Z_s = Z_e + (R_1 + R_2)$$

Example:

An installation has an external earth fault loop impedance of 0.5 Ω. Given that the lighting circuit has been wired in 1.5 mm^2 copper conductors having a resistance of 12.1 mΩ/m, and length 30 m, calculate the expected earth fault loop impedance for this circuit at an ambient temperature of 20 °C. For ambient temperatures other than 20 °C the value of $(R_1 + R_2)$ will need to be corrected (refer to Appendix H of this Guide).

$$(R_1 + R_2) = \text{length x resistance/m} = 30 \times 12.1/1000 = 0.363\ \Omega.$$

$$Z_s = Z_e + (R_1 + R_2) = 0.5 + 0.363$$

$$Z_s = 0.863\ \Omega$$

Note: For circuits rated at more than 100 A, the use of DC test current may lead to a value of Z_s which is optimistically low (due to the effects of circuit reactance) and may therefore be unacceptable. In such cases, an ohmmeter or continuity test instrument which supplies an AC test current at a frequency of 50 Hz should be used.

8.12 (iv) Verifying earth fault loop impedance test results

A circuit that relies on the protective measure ADS for fault protection is required to disconnect under fault conditions within the maximum time permitted by *BS 7671* for that type of circuit or location. Maximum disconnection times for final circuits are given in Table 41.1 (Regulation 411.3.2.2 refers) and for circuits not covered by Table 41.1 and distribution circuits, disconnection should be achieved in a time not exceeding 5 s in a TN system and 1 s in a TT system (Regulations 411.3.2.3 and 411.3.2.4 refer).

The maximum values of earth fault loop impedance to achieve the disconnection time vary with the different types of protective devices and also between manufacturers. Wherever possible the manufacturers' specific data should be used. However, in the absence of manufacturers' data, the values stated in the tables contained in Chapter 41 of *BS 7671* should be used to verify compliance (Appendix 3 of *BS 7671* refers). Tables 41.2, 41.3 and 41.4 of *BS 7671* give maximum permitted values of Z_s for different types of overcurrent protective devices. As shown in Fig 8.12 (iv), both the maximum permitted earth fault loop impedance and the measured value must be recorded on the 'Schedule of Circuit Details and Test Results' which forms part of NICEIC and ELECSA certificates and reports.

Maximum permitted Z_S for installed protective device*	Circuit impedances (Ω)					Insulation resistance			Polarity	Max. measured earth fault loop impedance, Zs
	Ring final circuits only (measured end to end)			All circuits (complete at least one column)		Live / Live	Live / Earth	Test voltage DC		
(Ω)	(Line) r_1	(Neutral) r_n	(cpc) r_2	$(R_1 + R_2)$	R_2	(MΩ)	(MΩ)	(V)	(✓)	(Ω)

Fig 8.12 (iv) Recording details in relation to Z_s

The remainder of this section looks at the verification of earth fault loop impedance test results where circuits are connected to a TN supply arrangement and also where circuits are connected to TT supply arrangement.

8.12(v) Verification of circuits connected to TN systems

Where an overcurrent protective device is used to provide fault protection, Regulation 411.4.4 requires that the characteristics of the device and the circuit impedance fulfil the following requirement:

$$Z_s \times I_a \leq U_0 \times C_{min}$$

where:

$\boldsymbol{Z_s}$ is the impedance in ohms (Ω) of the fault loop comprising:

- the source
- the line conductor up to the point of the fault, and
- the protective conductor between the point of the fault and the source.

$\boldsymbol{I_a}$ is the current in amperes (A) causing the automatic operation of the disconnecting device within the time specified in Regulation 411.3.2.2, or Regulation 411.3.2.3. Where an RCD is used this current is the residual operating current providing disconnection in the time specified in Regulation 411.3.2.2, or Regulation 411.3.2.3.

$\boldsymbol{U_0}$ is the nominal AC rms line voltage to Earth.

C_{min} is the minimum voltage factor to take account of voltage variations depending on time and place, changing of transformer taps and other considerations (stated in *BS 7671* as 0.95).

The tabulated values of maximum permitted earth fault loop impedances contained in Tables 41.2, 41.3 and 41.4 of *BS 7671* have been determined using the voltage factor C_{min} (0.95). Where a measured value of Z_s is obtained, it will need to be adjusted before a comparison is made with these tabulated values. This is because the values referred to in *BS 7671* are based on the conductors having been heated up by the passage of load current, which increases their resistance, whilst test results are usually obtained when the conductor temperature is somewhat lower. As a rule of thumb, the measured value of Z_s should not exceed 0.8 times the tabulated values given in *BS 7671* (see Appendix 3 of *BS 7671*).

Note: Other methods of correcting for the effects of temperature difference are not precluded.

For convenience, each pad of NICEIC and ELECSA certificates or reports contains a table that gives the limiting values of measured earth fault loop impedance when measured at ambient temperatures up to 20 °C. These limiting measured values are based on 80 % of the values given in *BS 7671* for 0.4 s and 5 s disconnection, as appropriate, rounded down. The boxes marked 'N/A' (Not Applicable) indicate either that the protective device is not commonly available or that, by virtue of its characteristics, the device is not generally appropriate for fault protection.

Limiting values of measured earth fault loop impedance for common overcurrent protective devices, for fault protection, operating at 230 V based on 80 % (approximately) of the values given in *BS 7671*

Rated current (A)	Fuses										Circuit-breakers to *BS 3871* or *BS EN 60898* or RCBOs to *BS EN 61009*					
	BS 88 (gG) Parts 2 and 6		*BS 1361* or *BS 1362*		*BS 3036*		*BS 88-2* Fuse systems E (bolted) and G (clip in)		*BS 88-3* Fuse system C		Type 1	Type 2	Type B	Types 3 & C	Type D	
	0.4 s	5 s	0.4 s	5 s	0.4 s	5 s	0.4 s	5 s	0.4 s	5 s	0.4 s & 5 s				0.4 s	5 s
2	N/A	N/A	N/A	N/A	N/A	N/A	26.48	34.96	N/A	N/A	N/A	N/A	N/A	N/A	N/A	N/A
3	N/A	N/A	12.46	17.63	N/A	N/A	N/A	N/A	N/A	N/A	N/A	N/A	11.64	N/A	N/A	N/A
4	N/A	N/A	N/A	N/A	N/A	N/A	12.46	16.64	N/A	N/A	N/A	N/A	N/A	N/A	N/A	N/A

Fig 8.12 (v) Limiting values of measured earth fault impedance for common overcurrent protective devices

8.12 (vi) Verification of circuits connected to TT systems

Where, disconnection is achieved by an overcurrent protective device and the protective equipotential bonding is in accordance with Regulation 411.3.1.2 the maximum disconnection times of Table 41.1 applicable to TN systems may be used (Regulation 411.3.2.2 refers).

For an overcurrent protective device used to provide fault protection the following condition specified in Regulation 411.5.4 shall be fulfilled:

$$Z_s \times I_a \leq U_0 \times C_{min}$$

Where an RCD is used to provide fault protection, as is normally the case, for circuit(s) connected to a TT system, Regulation 411.5.3 requires the following conditions to be fulfilled:

- The appropriate disconnection times of Regulation 411.3.2.2 or 411.3.2.4, and
- $R_A \times I_{\Delta n} \leq 50$ V

where:

R_A: is the sum of the resistances of the earth electrode and the protective conductor connecting it to the exposed-conductive-parts (in ohms). Where R_A is not known it may be replaced by Z_S.

$I_{\Delta n}$: is the rated residual operating current of the RCD.

The requirements of Regulation 411.5.3 are met if the Z_s of the final circuit protected by the RCD meets the requirements of Table 41.5 of *BS 7671* which are shown in the following table.

Table 8.5 Maximum earth fault loop impedance (Z_s) for non-delayed and time delayed 'S' Type RCDs to *BS EN 61008-1* and *BS EN 61009-1* for U_0 of 230 V

Rated residual current ($I_{\Delta n}$) mA	Maximum earth fault impedance Ω
30	1667
100	500
300	167
500	100

Note: the resistance of the installation earth electrode should be as low as practicable

8.13 Additional protection

The requirements for providing additional protection by RCDs or by supplementary protective equipotential bonding are detailed in Section 415 of *BS 7671*, and where additional protection is employed, the effectiveness of the measures applied should be verified by visual inspection and testing (Regulation 643.8 refers).

Additional protection by RCDs

RCDs with a rated residual operating current not exceeding 30 mA may be used to provide additional protection in the event of failure of basic and/or fault protection or carelessness by the user (Regulation 415.1.1 refers). Where this is the case each RCD should be inspected and tested to confirm that the relevant requirements of Chapter 41 of *BS 7671* are met. Suitable methods of verifying the effectiveness of RCDs used to provide additional or fault protection are outlined later in this Chapter (see 'Operation of RCDs').

Supplementary protective equipotential bonding

Supplementary equipotential bonding is generally used in locations or installations where there is an increased risk of electric shock. For example, supplementary equipotential bonding may be used in accordance with Regulation 415.2 to provide additional protection under fault conditions for a special location or installation covered by Part 7 of *BS 7671*. Where it is used, continuity testing as previously outlined in this Chapter is a suitable method of verifying compliance with Chapter 41 of *BS 7671*.

As detailed in Regulation 415.2.2, the resistance '*R*' between simultaneously accessible exposed-conductive-parts and extraneous-conductive-parts should satisfy the following, as appropriate:

- $R \leq 50\ V/I_a$ (AC systems)
- $R \leq 120\ V/I_a$ (DC systems)

Where, $\boldsymbol{I_a}$ is the current in amperes (A) causing operation of the protective device within the specified time or, for RCDs, $I_{\Delta n}$.

Note: The csa of supplementary bonding conductors should comply with the requirements of Regulation Group 544.2.

8.14 Operation of residual current devices

To verify the effectiveness of an RCD used to provide automatic disconnection in the event of a fault as required by Regulation 643.7.1 or to verify the effectiveness of an RCD used to provide additional protection as required by Regulation 643.8 the following inspections and tests are applicable:

- a visual inspection to confirm that the RCD is of an appropriate type, rated current or current setting, and sensitivity for its intended application,

- testing of the RCD using a suitable test instrument according to *BS EN 61557-6*, and
- operation of the integral test facility to confirm correct function.

An RCD may also be used for protection against fire. However, an RCD that satisfies the requirements for ADS is considered to satisfy the requirements for providing protection against fire.

Note: An RCD is not permitted to be used in a TN-C system.

8.14 (i) RCD testing

The purpose of the RCD test is to confirm that the device will operate within the disconnection time required by *BS 7671*.

Where an RCD is used to provide fault protection, the test should be performed at any current equal to or greater than the rated residual operating current of the device. The disconnection time should comply with the general requirements detailed in Chapter 41.

Where an RCD is used to provide additional protection, the test should be performed at any current equal to or greater than $5I_{\Delta n}$. The disconnection time should not exceed 40 ms (Regulation 643.8 refers).

Whilst product standard testing requires manufacturers' to subject RCDs to tests at half their rated residual operating current ($0.5I_{\Delta n}$), a half-rated test is not required by *BS 7671* to verify the effectiveness of an RCD, during initial verification or periodic testing.

Note: Testing RCDs at $0.5I_{\Delta n}$ might be useful when fault finding during maintenance or periodic inspection procedures.

8.14 (ii) Test procedure

Before an RCD test is conducted, the earth loop impedance of the circuit(s) protected by the RCD should be confirmed.

It should be noted that it is the RCD that is being subjected to a test and not the circuit or the load. Therefore, the instrument may be connected at any convenient point in the circuit(s) protected by the RCD, such as at a socket-outlet or at the outgoing terminals of the RCD. Any loads connected to the circuit(s) protected by the RCD should be disconnected for the purpose of the test.

Generally, the RCD test instrument will have a facility to indicate whether or not polarity is correct. Testing should only proceed where correct polarity is being indicated.

The test is performed for both the positive and negative half-cycles of the supply waveform and the longest operating time obtained should be recorded in the appropriate box on the 'Schedule of Circuit Details and Test Results'.

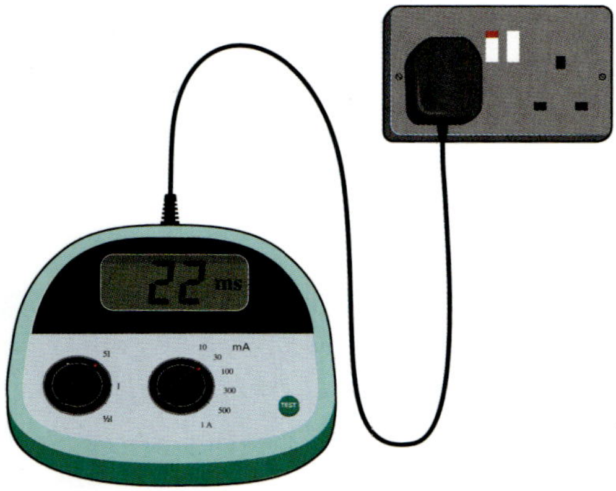

Fig 8.14 (ii) a Test performed at a socket-outlet

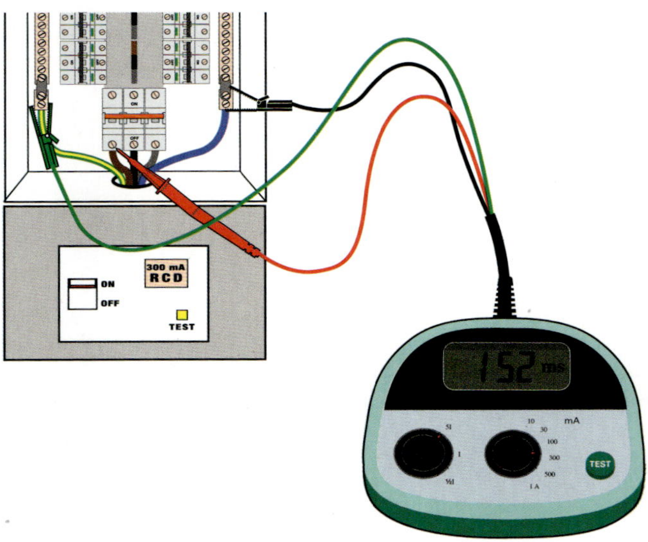

Fig 8.14 (ii) b Test performed at the load side of the RCD

8.14 (iii) Verification of the RCD test results

The results of RCD testing should be verified according to the function of the RCD. Where an RCD is installed to provide additional protection testing is required to verify that the RCD disconnects within 40 ms, as required by Regulation 415.1.1. Where an RCD is installed to provide fault protection, the testing should verify that the relevant disconnection times of Chapter 41 of *BS 7671* are met (Regulation 411.3.2.1 refers). If the disconnection time of Chapter 41 is not achieved when testing at $I_{\Delta n}$ the test should be repeated at a higher test current.

The following table provides a summary of the disconnection times required for fault protection.

Table 8.6 Disconnection times of AC circuits

Disconnection times of Chapter 41 of *BS 7671* for AC circuits (where 120 V < $U_0 \leq$ 230 V)			
Circuit type / application	**Earthing system**	**Disconnection time (seconds)**	***BS 7671* reference**
Final circuits with a rated current not exceeding 32 A	TN	0.4	411.3.2.2 Table 41.1
	TT	0.2	
Distribution circuits with rated current not exceed-ing 32 A	TN	5	411.3.2.3
	TT	1	411.3.2.4
Part of a reduced low voltage system	TN or TT	5	411.8.3

8.14 (iv) Operation of the integral test button

Each RCD should also be operated by means of its integral test facility. This manual test is required to confirm that the device is responding to its design level of sensitivity and that all the mechanical parts are functioning.

For the benefit of the user of the installation, a notice should be fixed at or near each RCD in the installation stating that the integral test button should be operated at intervals not exceeding six months (Regulation 514.12.2 refers).

> This installation, or part of it, is protected by a device which automatically switches off the supply if an earth fault develops. Test six-monthly by pressing the button marked 'T' or 'Test'. The device should switch off the supply and should then be switched on to restore the supply. If the device does not switch off the supply when the button is pressed, seek expert advice.

Fig 8.14 (iv) Six-monthly RCD test label

8.15 Verification of phase sequence

As required by Regulation 643.9, it should be verified that phase sequence has been maintained in polyphase circuits at all relevant points throughout the installation.

For initial verification, this is completed by the inspection and testing of conductors during and on completion of the installation. Before the supply is connected, continuity testing such as the (R_1+R_2) method undertaken to confirm polarity will also allow identification of both ends of the conductor to be verified and confirm that it is provided in accordance with Table 51 of *BS 7671*

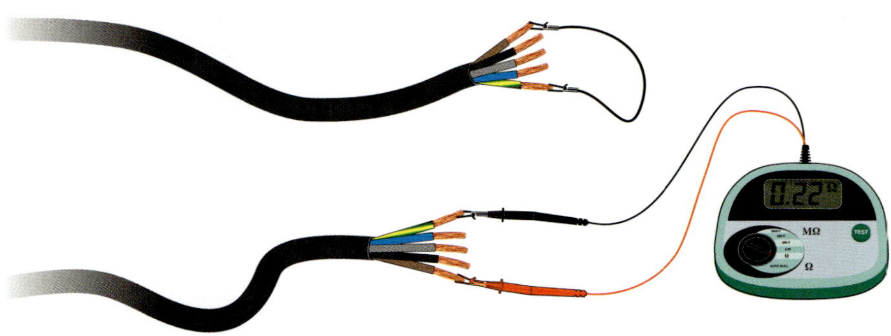

Fig 8.15 Verification of phase sequence

The continuity procedure shown should be repeated for each live (line and neutral) conductor of the circuit in turn. The procedure should be carried out for all distribution circuits throughout the installation.

A phase rotation instrument can be used to confirm that supplies to three-phase equipment such as rotating machines are in the correct sequence, so that when operated the machines will rotate in the intended direction.

8.16 Functional testing of assemblies

All equipment should be checked to ensure that it functions as intended, and has been properly mounted, adjusted and installed and that it operates correctly in accordance with the relevant requirements of *BS 7671* (Regulation 643.10 refers).

Functional testing of installed equipment includes switchgear and controlgear assemblies, drives and interlocks, motor drives and starters. Correct operation of protective devices and the effectiveness of any test facility incorporated in an RCD used to provide fault protection and/or additional protection should be verified.

In addition, the functional operation of any arc fault detection device (AFDD) connected to a circuit in accordance with Regulation 421.1.7 should be confirmed.

Where there is doubt as to the intended method of controlling any part of an installation (such as a control system), the installation designer should be consulted.

8.17 Verification of voltage drop

Verification of voltage drop is not normally required during the initial verification of either a new installation or an addition or alteration to an existing installation. However, where compliance with the voltage drop requirements of Section 525 of *BS 7671* needs to be verified this may be achieved by measurement or by calculation (Regulation 643.11 refers).

The requirements of Section 525 are deemed to be met if the voltage drop between the origin of the installation (usually the supply terminals) and a socket-outlet or the terminals of fixed current-using equipment does not exceed that stated in Section 6.4 of Appendix 4 of *BS 7671* (Regulation 525.202 refers).

For a low voltage installation supplied directly from a public low voltage distribution system, Section 6.4 of Appendix 4 gives maximum voltage drops of:

- 3 % for lighting circuits
- 5 % for other circuits.

Greater voltage drops than given in Section 6.4 may be accepted for a motor during starting periods and for other equipment with high in-rush currents provided it is verified that it is acceptable for the equipment (Regulation 525.203 refers).

8.17 (i) Evaluation of voltage drop by measurement

Suppose that it is desired to verify that the voltage drop for a 230 V, 50 Hz radial circuit supplying a 20 A lighting load (connected at the end of the circuit) does not exceed 5 %. Given that the circuit is connected directly to a distribution board at the origin of the installation, the conductors have thermoplastic (PVC) insulation and their resistance ($R_1 + R_n$) is 0.3 Ω when measured at an ambient temperature of 20 °C , the voltage drop can be verified as follows:

VD = measured value of $(R_1 + R_n) \times 1.2 \times I_b$

Where I_b is the design current of the circuit and 1.2 is a correction factor used to increase the conductor resistance from 20 °C (ambient) to 70 °C (the maximum permitted operating temperature for thermoplastic insulated conductors).

Therefore VD = 0.3 Ω x 1.2 x 20 A = 7.2 V

A voltage drop of 7.2 V is equivalent to 3.13 % which exceeds 3 % of 230 V and as such, the circuit voltage drop does **not** comply with *BS 7671*.

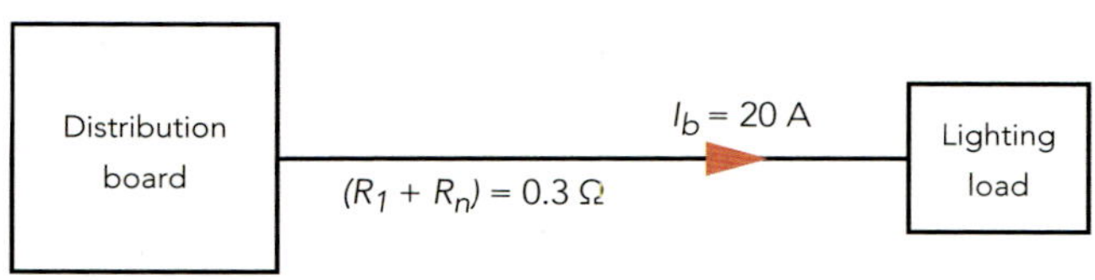

Fig 8.17 (i) Evaluation of voltage drop

Alternatively, the voltage drop for a circuit may be calculated by using the tabulated values of voltage drop given in Appendix 4 of *BS 7671*, in accordance with the instructions given in Section 6 of that appendix. To do this, it will be necessary to ascertain the type and size of cable by inspection, and to estimate the length of run for the circuit.

APPENDICES

Safe Isolation

All persons carrying out electrical installation or maintenance work must be able to carry out the safe isolation of all or part of an electrical installation.

Regulation 12 of the *Electricity at Work Regulations 1989* deals with means of cutting off the supply and for isolation. The regulation requires that where necessary to prevent danger, suitable means (including, where appropriate, methods of identifying circuits) are made available:

- for cutting off the supply of electrical energy to any electrical equipment, and
- for the isolation of electrical equipment.

Isolation is not simply the operation of an isolating device. It is a procedure, the objective of which is to ensure that the supply is cut off and remains cut off for reasons of safety. The process of isolation can be broken down into a number of steps as necessary for the particular circumstances. As a minimum, the following safe working procedure should be adopted before commencing inspection (*HSG85: Electricity at Work: Safe working practices* gives detailed guidance on devising safe working practices for people who carry out work on or near electrical equipment).

The example illustrated overleaf shows the minimum steps required to isolate the final circuits supplied by a single-phase consumer unit. The consumer unit includes an isolator and circuit-breakers.

Where circuits are protected by fuses enclosed in a consumer unit it may be necessary to isolate the whole consumer unit.

Step 1

Check it is safe and acceptable (with the occupier / user) to isolate. If the isolator is an off-load device, remove the load.

Open the means of isolation for the circuit(s) to be isolated and secure the isolating device in the open position with a lock or other suitable means.

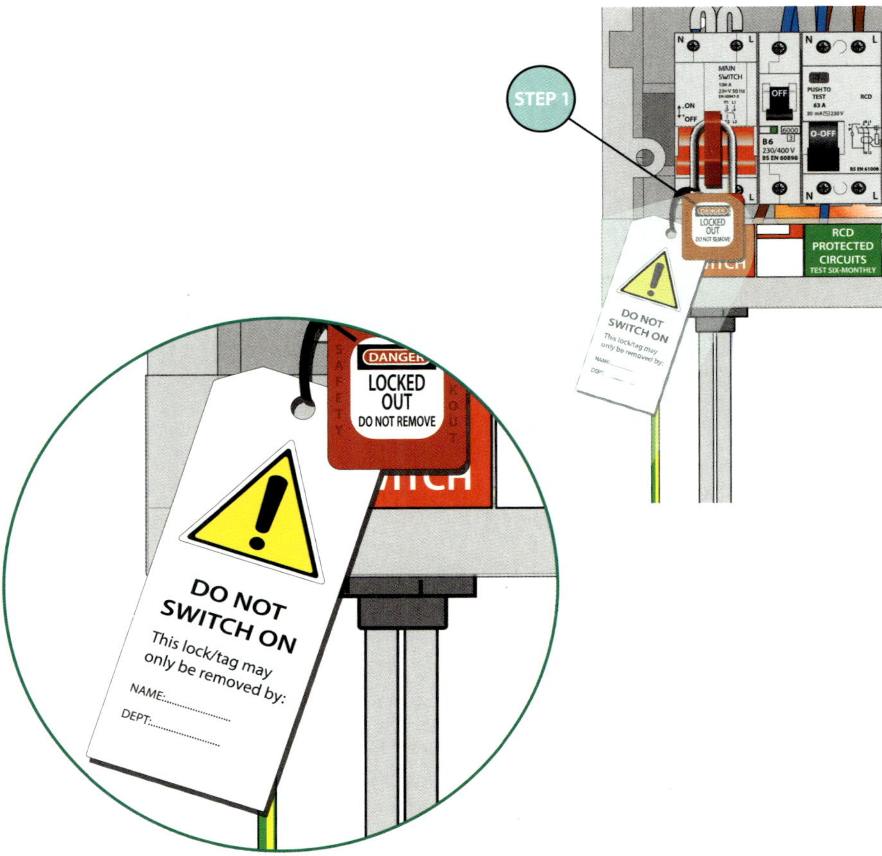

A label such as the one shown, should be attached to the padlock giving details of the person in charge of the key.

Step 2

Prove the correct operation of a suitable voltage detection instrument against a known voltage source. If the voltage tester does not have a built in test facility to verify its operation, a proving un t should preferably be used. Guidance on the use of electrical test instruments is given in Appendix B of this Guide.

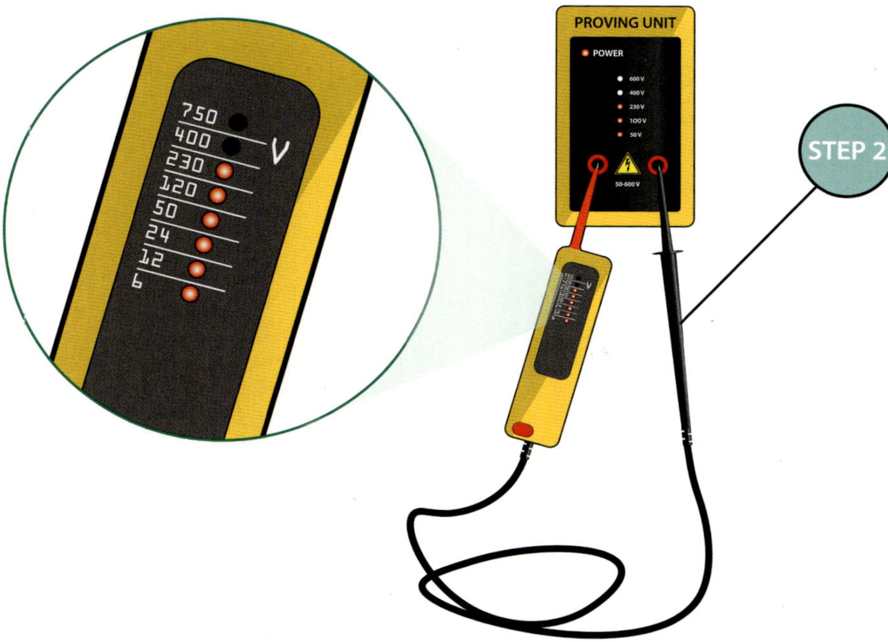

Step 3

Check that there is no dangerous voltage present on any circuit conductor to be worked on or near to. It is important to confirm that conductors are not energised, for example, due to a wiring fault. Check terminal voltages between: (i) line and earth, (ii) line and neutral (as shown) and (iii) neutral and earth.

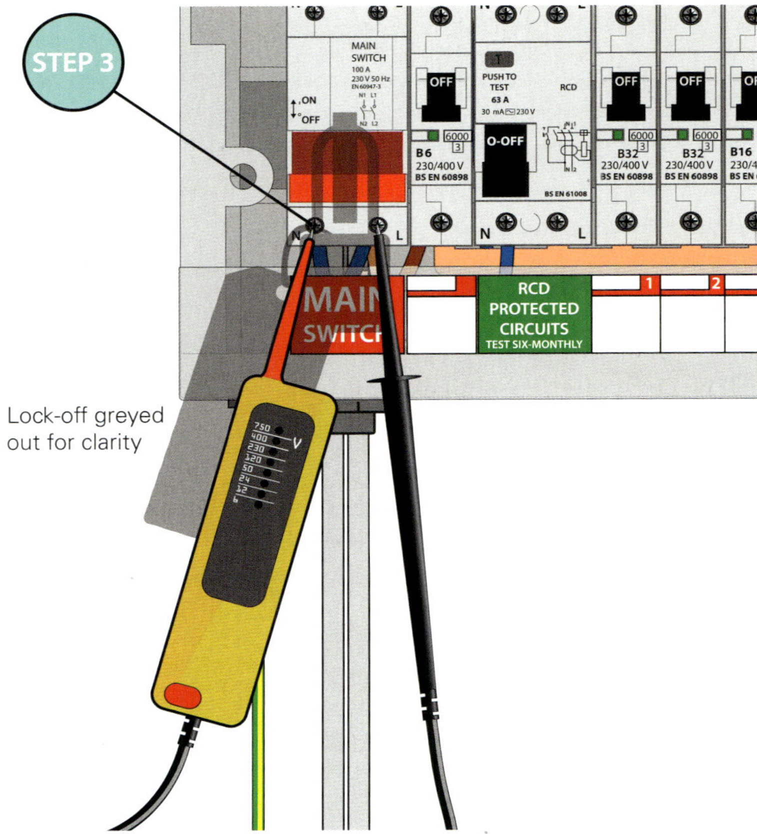

Step 4

Prove the voltage detection instrument again against the known source to check that it was functioning correctly when the circuit(s) was tested for the presence of voltage.

The Electrical Safety First Best Practice Guide 2, *Guidance on the management of electrical safety and safe isolation procedures for low voltage installations* has been produced by Electrical Safety First in conjunction with the Health and Safety Executive (HSE) and other organisations. Its purpose is to provide practical guidance for employers and the self-employed on safe isolation procedures to be followed when working on low voltage electrical installations.

The Guide can be downloaded from Electrical Safety First's website:

www.electricalsafetyfirst.org.uk

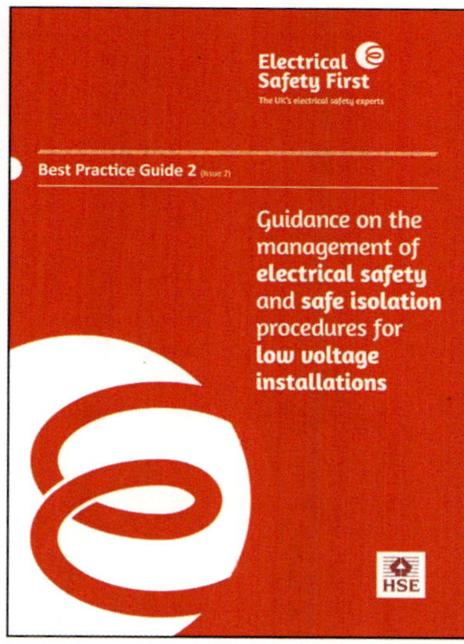

Guidance on the 'isolation procedure' is also provided in the NICEIC and ELECSA *Pocket Guide 5 – Guide to isolation procedure.*

All NICEIC and ELECSA Pocket Guides are available to download from the NICEIC and ELECSA website: **www.niceic.com and www.elecsa.co.uk**.

HSE Guidance Note *GS38: Electrical test equipment for use on low voltage electrical systems* gives advice on the selection and use of test equipment for circuits with rated voltages not exceeding 650 V. Those involved in electrical inspection and testing should familiarise themselves with the guidance given in this publication.

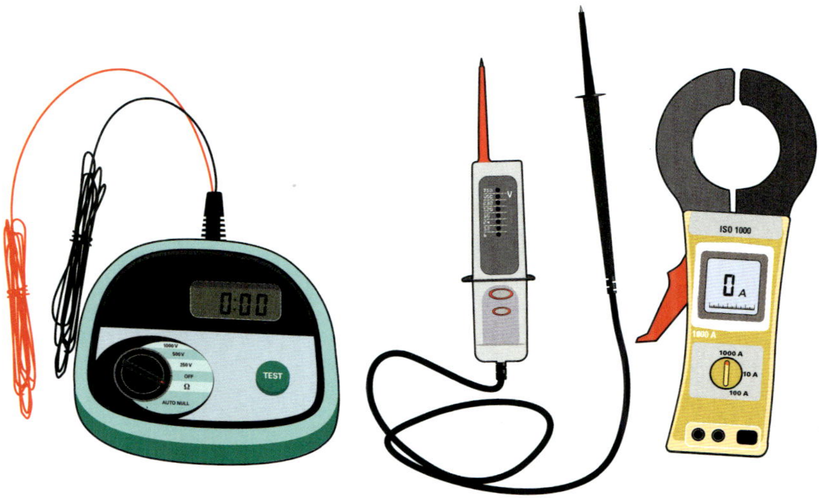

Aspects that should be checked with regard to test equipment include the following:

- Test instruments and their associated test leads must be suitably constructed and fit for the purpose for which they are to be used. Such equipment should preferably meet the safety requirements set out in *BS EN 61010* and the performance standards detailed in the relevant Part(s) of *BS EN 61557*.
- Test equipment should be checked for suitable rating for the anticipated voltage prior to testing any circuit which is, or may be, energised.
- Test instruments and their associated test leads should be checked before each use for signs of damage and deterioration, and be maintained such that they remain safe for use in the intended manner.
- Test instruments and test leads should be used only in the environment(s) for which they have been designed.

- In the case of voltage test lamps, the leads (and therefore the person using the instrument) should be protected by either: a high breaking capacity (HBC) fuse of rating not exceeding 500 mA, or a current limiting resistor and a fuse.
- In the case of voltage detectors, the current that may pass through the instrument is limited by the internal circuit design, protective impedance and current limiting resistors built into the probes.
- Probes should have no more than 4 mm of exposed metal tip. It is strongly recommended that the exposed metal tip does not exceed 2 mm.

GS38 can be downloaded from the health and safety website: **www.hse.gov.uk**

Note: Testing in a potentially explosive atmosphere should be carried out in accordance with the appropriate safety precautions of *BS EN 60079-17* (Regulation 643.1 refers).

The following table does not represent an exhausive list, but shows examples of the limited scope of the Minor Electrical Installation Works Certificate (MEIWC).

ACCEPTABLE USE OF THE MEIWC	
Examples of electrical work	**Comment**
Additional lighting point (luminaire and / or switching) on a single existing circuit.	The MEIWC may be used provided the existing circuit protective device is suitable for protection of the modified circuit, and other safety provisions are adequate.
Additional socket-outlet on a single existing radial or ring circuit.	The MEIWC may be used provided the existing circuit protective device is suitable for protection of the modified circuit, and other safety provisions are adequate.
Replacement of an accessory, such as a socket-outlet or a ceiling rose, on a like-for-like basis.	Although such replacement may be regarded as maintenance work, a record of the relevant safety tests undertaken may be recorded on the MEIWC, though other forms of recording such results are not precluded.
Replacement of an individual item of switchgear including a control switch, in a like-for-like manner for a single circuit.	This also may include the replacement of a single protective device.
Installation or upgrading of main protective bonding.	The MEIWC is suitable for such work.
Re-fixing, re-lidding existing wiring systems, or adding mechanical protection to existing equipment.	On condition that the circuit's protective measures and current-carrying capacity of conductors are unaffected.
NON-ACCEPTABLE USE OF THE MEIWC	
New circuit (lighting/power).	An electrical certificate (EIC or DEIC) must be issued for a new circuit.
New circuit (lighting, power etc.) connected to an existing protective device.	An electrical certificate (EIC or DEIC) must be issued for a new circuit.
Replacement of main switchgear, incorporating protective devices, for more than one circuit.	An electrical certificate (EIC or DEIC) must be issued for such work.
Modifications to more than one circuit.	A single MEIWC must not be used for modifications to more than circuit. (One MEIWC must be issued for each circuit altered or extended.
Portable appliance inspection and testing.	Such work is not within the scope of the MEIWC.

Recommended Initial Frequencies of Inspection of Electrical Installations

Type of installation	Maximum period between inspection and testing
General installations	
Domestic (owner occupied)	Change of occupancy / 10 years
Domestic (rented property)	Change of tenancy / 5 years
Commercial	Change of occupancy / 5 years
Hospitals	5 years
Offices	5 years
Industrial	3 years
Buildings open to the public	
Cinemas	1-3 years
Leisure complexes excluding swimming pools	3 years
Restaurants and hotels	3 years
Theatres	3 years
Special installations	
Agricultural & horticultural	3 years
Caravans (infrequently moved)	3 years
Caravans (frequently moved)	1 year
Caravan parks	1 year
Highway power supplies	6 years
Marinas	1 year
Emergency lighting	3 years
Fire alarms	1 year
Launderettes	1 year
Petrol filling stations	1 year
Construction site installations	3 months

The next inspection date should take into consideration any legislative or licensing requirements and the frequency and quality of maintenance that the installation can reasonably be expected to receive during its intended life.

An extraneous-conductive-part is defined in *BS 7671*, as a part that is conductive, and is:
liable to introduce a potential, generally Earth potential, but does not form part of the electrical installation.

It is now commonplace for installation pipework to be formed of non-conductive material, and as such the pipework would not be deemed an extraneous-conductive-part and require connecting to the MET. Furthermore, where a metallic pipe is comprised of an insulating section on entry into the building it is not deemed an extraneous-conductive-part (Regulation 411.3.1.2 refers). It should also be recognised that an item that is conductive may not necessarily introduce a potential. In order to present a risk of electric shock the item must be at Earth potential (that is, in contact with the general mass of the Earth) and, crucially, it must also be accessible. A conductive part, such as a metal duct or pipe, that is covered with, or enclosed by, insulating material, or an item that is out of reach and not in contact with any exposed-conductive-part or extraneous-conductive-part is not liable to introduce a potential.

For a new installation, any doubt regarding protective bonding should be referred to the installation designer for consideration and confirmation as to the requirement for main, and where applicable, supplementary bonding. For an inspection of an existing installation, the inspector should make an assessment based on all the relevant circumstances. One method of verifying whether a conductive part requires bonding is to carry out insulation resistance testing at 500 V DC between the MET and the metallic part, or pipework.

Where the measured value of insulation resistance between the item and the MET is such that it would prevent any dangerous level of current flow between the part and Earth, the item may be considered to be an isolated metallic part rather than an extraneous-conductive-part. Based on the 30 mA (RCD) current value (the level of current that *BS 7671* deems a shock risk), and a body resistance of 1000 Ω for a 230 V (50 Hz) supply, the metallic part would not require connecting to the MET where the measured value of insulation resistance is greater than:

$$\frac{230\ (\text{V})}{0.03\ (\text{A})} - 1000\ (\Omega) = 6.67\ \text{k}\Omega$$

Note: The value of 1000 Ω is chosen from *PD IEC/TS 60479-1* (based on hand-to-hand contact in dry conditions).

For example, for the insulation resistance test shown in Fig E, the test reading of 5 megohm is sufficient to deem the part as isolated metalwork rather than an extraneous-conductive-part. On the basis of this test result the installation of a protective bonding conductor between the pipe and the MET is not required.

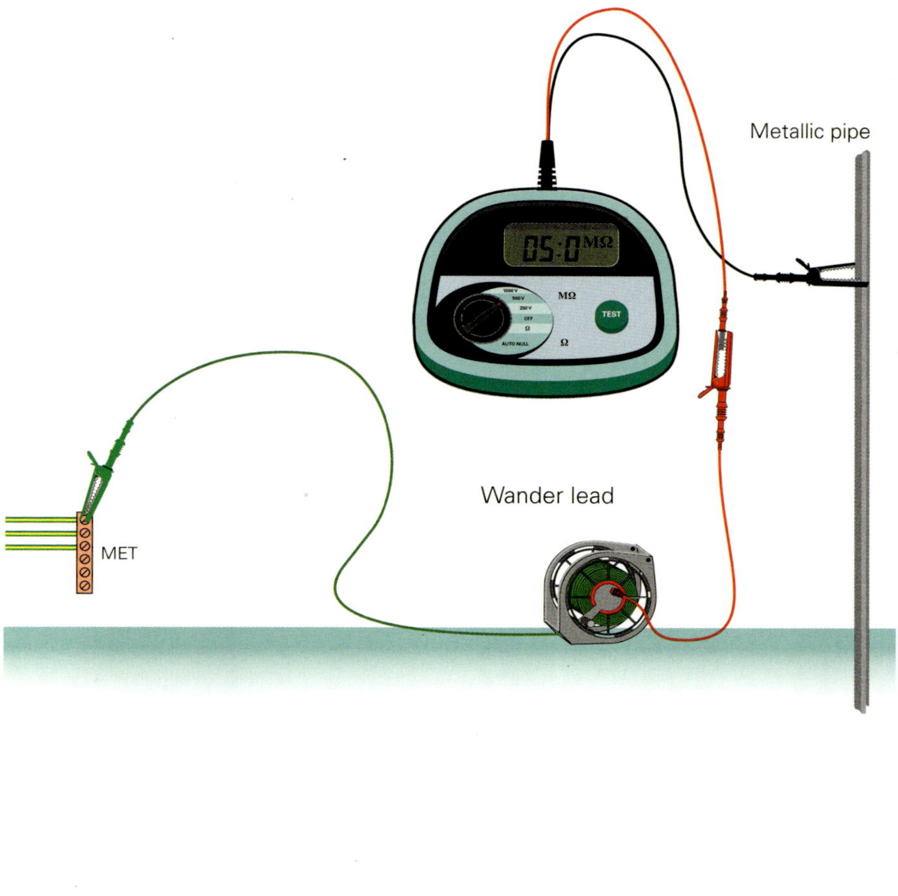

Fig E Insulation resistance testing to verifying whether protective bonding is required

International Protection (IP) Code

FIRST NUMERAL* & **		SECOND NUMERAL***	
No.	**DEGREE OF PROTECTION**	**No.**	**DEGREE OF PROTECTION**
0	* No special protection. ** No special protection.	0	No special protection.
1	* Protection against accidental or inadvertent contact by a large surface of the body, e.g. hand, but not against deliberate access. ** Protection against ingress of large solid objects greater than or equal to 50mm diameter.	1	Protection against drops of water. Drops of water falling on enclosure shall have no harmful effect.
2	* Protection against contact by standard finger. ** Protection against ingress of medium size bodies greater than 12.5 mm diameter.	2	Protection against water. Dripping water on enclosure shall have no harmful effect when the enclosure is tilted at any angle up to 15° from the vertical.
3	* Protection against contact by tools more than 2.5 mm diameter. ** Protection against ingress of solid objects greater than 2.5 mm diameter.	3	Protection against spraying water. Water sprayed against enclosure at any angle up to 60° from vertical shall have no harmful effect.
4	* Protection against contact by wires of 1 mm diameter. ** Protection against ingress of solid objects greater than 1.0 mm diameter.	4	Protection against splashing water. Water splashed against enclosure from any direction shall have no harmful effect.
5	* Protection against contact by wires of 1 mm diameter. ** DUST-PROTECTED – Protection against harmful deposits of dust. Dust may enter but not in amount sufficient to interfere with satisfactory operation.	5	Protection against water jets. Water projected in jets against enclosure from any direction shall have no harmful effect.
6	* Protection against contact by wires of 1 mm diameter. ** DUST-TIGHT – No ingress of dust.	6	Protection against powerful water jets. Water projected in powerful jets against enclosure from any direction shall have no harmful effect.
		7	Protection against temporary immersion in water. It shall not be possible for water to enter the enclosure under stated conditions of pressure and time.
		8	Protection against continuous immersion in water. It shall not be possible for water to enter the enclosure, to harmful effect, under stated conditions of pressure and time.
		9	Protection against high pressure & temperature water jets. Water projected at high pressure & temperature against enclosure from any direction shall have no harmful effect.

IP CODE NOTES

* PROTECTION OF PERSONS AGAINST CONTACT WITH HAZARDOUS PARTS INSIDE ENCLOSURE

** PROTECTION OF EQUIPMENT INSIDE ENCLOSURE AGAINST INGRESS OF SOLID BODIES

*** PROTECTION OF EQUIPMENT INSIDE ENCLOSURE AGAINST INGRESS OF WATER

- Degree of protection is stated in form IPXX.
- Protection against contact or ingress of water respectively is specified by replacing first or second X by digit number tabled, e.g. IP2X defines an enclosure giving protection against finger contact but without any specific protection against ingress of water or liquid.

SUPPLEMENTARY LETTERS

The IP code may also include a supplementary letter after the second numeral indicating:

H High-voltage.
M Tested for harmful effects due to the ingress of water when moving parts of equipment are in motion.
S Tested for harmful effects due to the ingress of water when moving parts of equipment are stationary.
W Suitable for use under specified weather conditions and provided with additional protective features or processes.

Refer to *BS EN 60529:1992*, as amended, for full information on degrees of protection provided by enclosures.

Code C1 (Danger present)
Observations that would almost certainly warrant a Classification code C1 include:

- Exposed live parts that are accessible to touch, such as where:
  - a fuse carrier or circuit-breaker is missing from a distribution board and a blanking piece is not fitted in its place
  - terminations or connections have no (or damaged) barriers or enclosures
  - live conductors have no (or damaged) insulation
  - an accessory is badly damaged
- Conductive parts that have become live as the result of a fault
- Incorrect polarity

Code C2 (Potentially dangerous)
Observations that would usually warrant a Classification code C2 include:

- Absence of an effective means of earthing for the installation
- A public utility water pipe being used as the means of earthing for the installation
- A gas or oil pipe being used as the means of earthing for the installation
- Cross-sectional area of the earthing conductor does not satisfy adiabatic requirements (that is, does not comply with Regulation 543.1.1)
- Absence of a circuit protective conductor for a lighting circuit supplying one or more items of Class I equipment, or connected to switches having metallic face plates
- Absence of a circuit protective conductor for a circuit, supplying one or more items of Class I equipment
- Absence of earthing at a socket-outlet
- Absence of main protective bonding
- Inadequate cross-sectional area of a main protective bonding conductor where the conductor is less than 6 mm^2 or where there is evidence of thermal damage
- Absence of supplementary bonding where required

Code C3 (Improvement recommended)
Observations that would usually warrant a Classification code C3 include:

- Absence of RCD protection for a socket-outlet that is unlikely to supply portable or mobile equipment for use outdoors, does not serve a location containing a bath or shower, and the use of which is otherwise not considered by the inspector to result in potential danger. (Note: code C2 would apply if the circuit supplied a socket-outlet in a location containing a bath or shower in accordance with Regulation 701.512.3)
- Absence of RCD protection for cables installed at a depth of less than 50 mm from a surface of a wall or partition where the cables do not incorporate an earthed metallic covering, are not enclosed in earthed metalwork, or are not mechanically protected against penetration by nails and the like
- Reliance on a voltage-operated earth-leakage circuit-breaker for fault protection (protection against indirect contact), subject to the device being proved to operate correctly. (If the circuit-breaker relies on a water pipe not permitted by Regulation 542.2.6 as the means of earthing, this would attract a Classification code C2.)
- Absence of RCD test notice
- Absence of circuit protective conductors in circuits having only Class II (or all-insulated) luminaires and switches
- Absence of 'Safety Electrical Connection - Do Not Remove' notice
- Sheath of an insulated and sheathed non-armoured cable not taken inside the enclosure of an accessory, such as at a socket-outlet or lighting switch. (Note: code C2 would apply if unsheathed cores are accessible to touch and / or likely to come into contact with metalwork)
- Bare protective conductor of an insulated and sheathed cable not sleeved with insulation, colour coded to indicate its function
- Electrical equipment having an inadequate degree of ingress protection (IP rating) for the external influences likely to occur in the location, if this does not result in potential danger
- Socket-outlet mounted in such a position as to result in potential damage to socket, plug and/or flex
- Absence of a notice indicating that the installation has wiring colours to two versions of *BS 7671* (if appropriate)
- Absence of circuit identification details

Note: Non-compliances with the requirements of the current edition of *BS 7671* that do not give rise to danger or need improvement do not need to be recorded on the Report.

The following Table gives resistance per metre values of copper conductors of single and multicore cables, whether of solid or stranded construction, at 20 °C, in units of milliohms per metre (mΩ per m).

The table is based on data contained in Tables 1 and 2 of *BS EN 60228*[1]

Table H1 – Maximum resistance of copper conductors at 20 °C

Conductor cross-sectional area (mm^2)	Resistance per metre of copper conductor (mΩ per m)
1	18.1
1.5	12.1
2.5	7.41
4	4.61
6	3.08
10	1.83
16	1.15
25	0.727
35	0.524
50	0.387

[1] *BS EN 60228:2005 Conductors of insulated cables*

Conductor resistance values at a conductor temperature other than 20 °C

Resistance values shown in Table H1 may be corrected to other resistance values, for conductor temperatures other than 20 °C, by using the formula:

$$R_x = R_{20} (1 + (0.004 \times (T_x - T_{20})))$$

Where:

R_x	is the corrected conductor resistance value, in mΩ, at the required temparature x °C
R_{20}	is the conductor resistance value at 20 °C, in mΩ, (see Table H1 for values)
0.004	is the approximate resistance temperature coefficient for copper
T_x	is the required x °C conductor temperature
T_{20}	is the 20 °C conductor temperature.

Calculation of conductor resistance at 30 °C

To calculate the resistance of a 6 mm^2 cross-sectional area (csa) copper conductor at a temperature of 30 °C:

T_{30} = 30 °C

R_{20} = 3.08 mΩ for a 6 mm^2 csa copper conductor, at 20 °C. (The value of 3.08 mΩ is taken from column 2 of Table H1.)

Substituting the above values in the formula gives:

$$R_{30} = R_{20} (1 + (0.004 \times (T_{30} - T_{20})))$$

$$R_{30} = 3.08 (1 + (0.004 \times (30-20))) = 3.08 (1 + (0.004 \times 10))$$

$$R_{30} = 3.08 \times (1 + 0.04)$$

$$R_{30} = 3.08 \times 1.04$$

$$R_{30} = 3.2 \text{ m}\Omega$$

It can be seen from the above calculation that the original value of conductor resistance at 20 °C is increased by a multiplication factor of 1.04 for a conductor temperature of 30 °C. (30 °C is the 'assumed initial conductor temperature' used in Table 54.2 of *BS 7671*.)

Conductor resistance at 70 °C

By carrying out a calculation similar to that shown in the previous exercise, it can be shown that the resistance of a conductor at 70 °C is 1.2 times its resistance value at 20 °C. (70 °C is the 'assumed initial conductor temperature' used in Table 54.3 of *BS 7671* for 70 °C thermoplastic insulated cables.)

These commonly used conductor temperature correction factors are given in Table H2, for ease of reference.

Further information on conductor temperature correction multiplication factors

Temperature correction factors for conductor resistance to correct measured resistance t °C are given in Table A.1 of *BS EN 60228* and exact formulae for temperature correction factors are stated in Annex B of that standard.

Table H2 – Resistance temperature multiplication correction factors for protective conductors with thermoplastic insulation, which are protected by standard overcurrent devices (given in Tables 41.2, 41.3 and 41.4 of *BS 7671*)

Protective conductor with thermoplastic (PVC) insulation	Initial conductor temperature (see Tables 54.2 and 54.3 of *BS 7671*)	20 °C conductor resistance correction multiplication factor (for applying to values in Table H1)
Applies where conductor is not incorporated in a cable or bunched with cables or bare conductors in contact with cable covers	30 °C	1.04
Applies where conductor is incorporated in a cable or bunched with other cables	70 °C	1.2